DISEASES OF GREENGRAM

AND

BLACKGRAM

S.C. AGRAWAL
Department of Plant Pathology,
J.N. Krishi Vishwa Vidyalaya,
R.A.K. College of Agriculture,
SEHORE-466001 (M.P.)

INTERNATIONAL BOOK DISTRIBUTORS
9/3, RAJPUR ROAD, DEHRA DUN
(INDIA)

Published by:
International Book Distributors
DEHRA DUN (India).

I.S.B.N. : 81-7089-161-2

Published by **R.P.S. Gahlot** for
International Book Distributors, Dehra Dun
Computer Typeset by Modern Computer System, Dehra Dun
Printed at Valley Offset Printers and Publishers
Ph: 27497, Fax: 91-135-27497
15-B, Rajpur Road Dehradun-248001

PREFACE

A considerable amount of research work has been done on diseases of greengram and blackgram during last thirty years and a lot has to be done. The information generated is lying scattered and is not available to the scientists working on this aspect. There has been a long felt urgent need of compiling this information for accelerating the research and saving these important pulse crops from diseases.

This book is a comprehensive and critical review of the world literature published on this aspect during 1918 to 1990. Diseases caused by various agencies are dealt in different chapters. A separate chapter has been included on seed pathology and storage. The information available on nematode diseases is very scanty. On virus diseases large number of publications have been made but in most cases incomplete study has been reported which creates lot of confusion. Many viruses have been mentioned as new reports without any proper identifications. Such viruses have been dealt in this book, as such. Future planning of research work on viruses needs careful thinking so as to avoid further confusion. This reference book primarily intended for Plant Pathologists involved in research on greengram and blackgram, will also be useful to teachers, post graduate students of Plant Pathology and the extension specialists interested in the development of pulses.

I wish to acknowledge my gratefulness to the various authors of scientific publications which form the basis of this book. The book includes several photographs depicting symptoms of important diseases few of which were contributed by Dr. K. Abbaiah, Lam, Dr. R. A. Singh, Kanpur and Dr. D.J. Patel, Anand. I am thankful to them. My heart-felt thanks are due to Dr. D.K. Sharma, Vice chancellor, Dr. D.P. Nema, Director Research Services, JKNVV, Jabalpur and Dr. P.K.R. Nair, Dean Agriculture College, Sehore, for encouragement, cooperation and help in various ways. I am thankful to my colleagues Drs. S.K. Srivastava, K.V.V. Prasad, R.K. Gupta and specially Miss Sushma Nema for their help.

I highly appreciate my youngest daughter Vaishali who prepared all the line drawings used in this book and my eldest son Arvind for his help in proof correction.

With a word of thanks, I would like to express my deep sense of gratitude to my wife Mrs. Leela Agrawal, and sons Anil and Neeraj whose cooperation and encouragement enabled me to complete this work within a short period of time.

I also thank the publishers, M/S International Book Distributors Dehra Dun for undertaking the task of bringing out the book with great interest.

Sehore
May 11, 1991

S.C. AGRAWAL
Department of Plant Pathology
J.N. Krishi Vishwa Vidyalaya,
R.A.K. College of Agriculture,
SEHORE- 466001 (M.P.)

TABLE OF CONTENTS

CHAPTER 1
INTRODUCTION

Greengram, Goldengram or Mungbean (*Vigna radiata* (L) Wilczek) and Blackgram, Blackbean or Urdbean (*Vigna mungo* (L) Hepper) formerly known as *Phaseolus aureus* Roxb. and *Phaseolus mungo* L. respectively are important pulse crops of Southern Asia. Some taxonomists consider them belonging to the same species *Vigna radiata* (L) Wilczek and maintain their identity by retaining sub species *aureus* and *mungo* for greengram and blackgram respectively.

Both are the good sources of protein in the diet of vegetarians in the form of *Dal.* Greengram is used as sprout also and grown widely in India, Pakistan, Sri Lanka, Thailand, Laos, Combodia, Vietnam, Indonesia, Malaysia, South China, Formosa, Philippines, Taiwan etc. It is also grown to a lesser extent in many parts of the Africa, United States ofAmericab and Australia.The cultivation of the blackgram is restricted to wet tropics particularly India, Pakistan, Sri Lanka, Myamnar and parts of South-east Asia, Africa and America. In West Indies, it is grown mainly as a green manure crop under the name Wooly pyrol (3). In India, it is also known as *Mash*.

In India, greengram and blackgram are grown on about 31.3 and 28.6 lakh hactrares with a total production of 12.4 and 10.5 lakh tonnes respectively (6). Thus the average production of these crops comes to less than 0.4 tonnes/ha as against their yield potential of about 1.5 to 2 tonnes/ha. The important reason of such a low productivity of these crops is their susceptibility to various diseases caused by fungi, bacteria, viruses, nematodes etc. In India, they are grown mainly during the rainy season (*Kharif*) when the environmental conditions make them more vulnerable to the attack of diseases like yellow mosaic, cercospora leaf spot, bacterial blight etc. causing heavy losses in yield. It is estimated that the yellow mosaic alone can cause 30-70 per cent losses in yield (5). Powdery mildew is the very serious problem of these crops in some of the areas particularly on winter crop. However, these crops are less vulnerable during the summer season but its scope is limited due to non availability of irrigation in summer. During the last decade the increased irrigation sources resulted in more than 250% increase in the area under summer mung in India (5).

Although, these crops have been suffering a lot due to various diseases, the research work on this aspect has been neglected in the past. Only few fungal diseases of these crops were described in India Up to 1918(4). During the period of 1918 to 1960, only 80 reports were written on this aspect and most of them were unpublished (1). Research work on diseases of greengram and blackgram started in India after 1960 and 90 papers were published up to 1970. The research on pulses was strengthened with the start of All India Coordinated Project on Pulses Improvement in late sixtees resulting in 370 and 350 research publications on diseases of these crops during 1970-1980 and 1981-89 respectively (1). A good research programme on greengram was started and still running at Asian Vegetable Research and Development Centre, Taiwan where two International Mungbean Symposia were organized in 1977 and 1987 in which research achievements on these crops were reviewed.

The literature on diseases of pulses including greengram and blackgram has been reviewed by several workers (2, 6, 7, 8, 9). Most of these reviews covered the deseases of all the pulses and hence, desired attention was not paid to these two neglected but important crops. Much of the important information generated on this aspect is still scattered. In this book, efforts have been made to review complete literature on this subject published so far in the world.

REFERENCES

1. Agrawal,S.C. 1989. Diseases of blackgram and greengram (An annotated bibliography 1918-1988). *R.A.K. Agriculture College, Sehore, Madhya Pradesh, India,* pp. 229 + XXI

2. Amin, K.S. and Singh, R.A. 1987. Diseases of mung, urid and pea and their control. In: "Plant Protection in Field Crops" (M. Veer Bhadra Rao and S.Sithananthan eds.), *Plant protection Association of India*, 203-217 pp.

3. Baldev, B. 1989. Origin, Distribution, Taxonomy and Morphology. In: "Pulse Crops" (B. Baldev, S. Ramanujam and H.K. Jain eds.), *Oxford and IBH Publishing Co. Pvt. Ltd. 3-51* pp.

4. Butler, E.J. 1918. "Fungi and Diseases in Plants". *Thacker Spink and Co. Calcutta*, 547 p.

5. Chandra, S. 1984. Increasing Pulse Production in India, Constraints and Opportunities. In: "Pulse production, Constraints and Opportunities". (H.C. Srivastava, S. Bhaskaran, K.K.G. Menon, S. Ramanujam and M.V. Rao eds.). *Oxford and IBH Publishing Co. Pvt. LTD.* 24-39 pp.

6. Grewal, J.S. 1988. Diseases of Pulses crops - an overview. *Indian Phytopathology* 41 : 1-14.

7. PANS 1981. "Pest Control in Tropical Grain Legumes." *PANS Mannual* No. 5 : 206 p.

8. Sen Gupta, P.K. 1974. Diseases of major pulse crops in India. *PANS* 20 : 409-415.

9. Singh, R.A., Gurha S.N. and Sahambi, H.S. 1981. Diseases of kharif pulses. *Indian Farming* (Special issue on pulses) August 1981.

CHAPTER 2
FUNGAL DISEASES

More than 35 diseases of greengram and blackgram, caused by fungi, have been reported but only few of them are economically important and occur widely. They are : powdery mildew, cercospora leaf spot, macrophomina disease and anthracnose. Some of the diseases like, mungbean scab, rust, false rust, angular black spot are capable of damaging the crop but they are restricted in certain regions/country only. Many of the foliar diseases e.g. leaf spots caused by species of *Heterosporium, Dactuliophora, Drechslera, Ascotricha, Alternaria, Septoria, Phyllosticta/Phoma* etc. have been observed rarely or not at all after their first report. The fungal diseases of greengram and blackgram can be classified in four categories.

Root and stem diseases : They are caused by the pathogenes which mainly infect root, collar or stem portion of the plant producing damping off, root rot, wilt, collar rot or stem rot symptoms. Important pathogens involved in this group of diseases are species of *Fusarium, Pythium, Cephalosporium, Thielaviopsis, Phytophthora* etc.

Multiphase diseases : Some pathogens produce two or more distinctly different phases of symptoms. *Macrophomina phaseolina* and *Rhizoctonia solani* are mainly soil-borne in nature causing seedling blight, root rot or stem rot symptoms. Under certain environmental conditions, they cause serious foliage blight known as macrophomina leaf blight and web blight respectively. In this phase the pathogen spreads very fast through air or beating rain involving large area of the crop. this phase of the disease may not appear in a particular season. At late stages even pods may be infected.

Foliar diseases : The pathogens responsible for the foliar diseases of greengram and blackgram spread mainly through the air infecting one or more of the above ground parts of the plant. Powdery mildew can infect almost all the plant parts while angular black spots are seen on leaves only.

Pod diseases : Large number of pathogens causing foliar diseases may also infect the pods but there are certain pathogens which are reponsible for pod rot only under saturated atmospheric conditions. They are not reported to infect other parts of plants of these crops under normal conditions. Important are diplodia pod rot, pythium pod rot etc.

These diseases are being described in this chapter in detail.

2.1. DAMPING OFF, SEEDLING MORTALITY AND STEM ROT :

The disease is widely distributed (CMI description map No. 309) on several leguminous crops. On greengram and blackgram, the disease was first reported from Iran (176). later, it was reported from India (262), Malaysia (3) and Philippines (129). In India, 8-12% damping off was reported on blackgram under wet-land conditions (123) while under artificially inoculated condition 96.4 and 70% death occured at pre- and post-emergence stages respectively. The greengram is attacked by many of the species of *Pythium* resulting in heavy losses (314).

Symptoms :

Raghunathan *et al.* (262) described damping off symptoms on greengram caused by *Pythium indicum* in India. Light brown, water soaked lesions appear near the hypocotyl region which spread towards root. The root becomes yellowish to dark brown and is constricted at the ground level. Immediate collapse of infected seedling is noticed (Fig. 2.1).

Fig. 2.1
Damping off in greengram

Pythium aphanidermatum is the most common pathogen causing damping off in greengram as well as blackgram in India and Iran. Infection occurs at the ground level which spreads rapidly up te hypocotyl giving it a grey-green wet appearance and the seedlings collapse (231). In adult plants, the disease is characterized by a grey-green, water soaked girdle of the stem which expands from soil level up to the and occasionally includes parts of the lower branches. Leaves and pods are also attacked. The infected area is soft and slimy to the touch. The stem cortex is easily stripped off and whitish, cottony fungal growth occurs at the base of the stem during period of high humidity. infercted plants rapidly wilt and die after defoliation. When the node is infectd, the fungus moves into the leaf, petiole, stem, branches and main stem. The leaf petiole is readily attacked and killed. Infected leaves are blighted and fall off prematurely. Flowers and pod infections lead to abscission of the flower and pod-rot. Other fungi including *Myrothecium roridum, Colletotrichum capsici* often colonize the infected area on death of the plant.

Causal organisms :

Pythium aphanidermatum (Edson) Fitz.
Pythium indicum Balkr.
Pythium ultimum Trow.
Pythium debaryanum Hesse.
Pythium splendens var. *hawaiianum*

Out of these *P. aphanidermatum* is most common on greengram and blackgram (132). The isolate of this fungus from blackgram produces cottony white growth of coenocytic *mycelium* on oat meal agar. Lobed *sporangia* are produced by forming a septum at the base on hyphal tip. The *antheridium* is clavate and of paragynous type while the *oogonium* is spherical and aplerotic ranging from 15.6-31.2μ in diameter. The identity of the species is based on its irregularly lobed *sporagia* and paragynous *antheridia* (132). According to Walker (453) the sporangia are, inflated, filamentous, branched or unbranched; *zoospores* 7.5 x 12.0 μ; oogonia, spherical, terminal, 22 x 27 μ; *antheridia*, monoclinous or diclinous, intercalary or terminal, 1 or 2 per oogonium, dome-shaped, 9-11 x 10-14 μ; *oospores* aplerotic, single, 17-19μ in size (Fig. 2.2).

The pathogen is reported to attack greengram in Iran (176). The virulence of the pathogen is increased by addition of nutrition to the

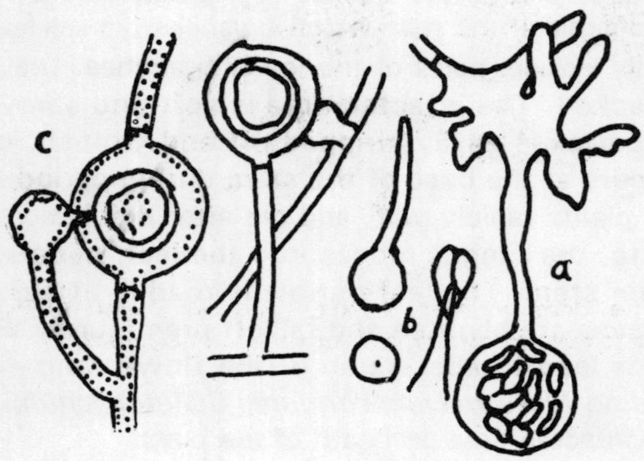

Fig.2.2
Pythium aphanidermatum : a. Sporangium, b. Flagillate, encysted and germinating zoospores, c. Oogonia with antheridia.

inoculum particularly those supplyig nitrogen sources utilized by the fungus. The virulence is also increased by exudates obtained from the germinating seeds of greengram (177). The exudates obtained at 12 and 42°C are more effective as compared to those obtained at intermediate temperature, because of the presence of higher sugars and aminoacid contents. It has been mentioned that at low inoculum density, favourable source of nitrogen supply is necessary for infection of greengram seedlings (178).

P. indicum is reported to cause damping off of greengram in India (262). However, the pathogen is not described in detail.

P. ultimum (fig. 2.3) is reported to infect mungbean seedlings in Iran (176) and its virulence is reported to increase by addition of nitrogen sources to the inoculum.

P. debaryanum (fig. 2.4) isolated from groundnut roots is reported to infect blackgram in India (263) but it has not been reported to occur under natural conditions on greengram or blackgram.

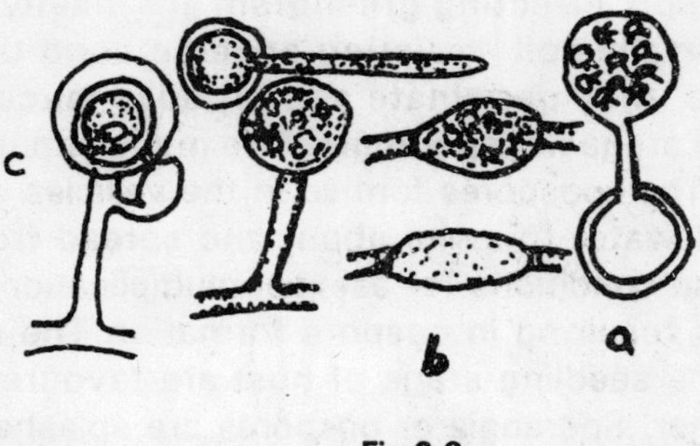

Fig.2.3
Pythium ultimum : a. Oospore forming vesicle, b. Intercalary sporangia which germinate forming germ tube, c. Oogonium and antheridium.

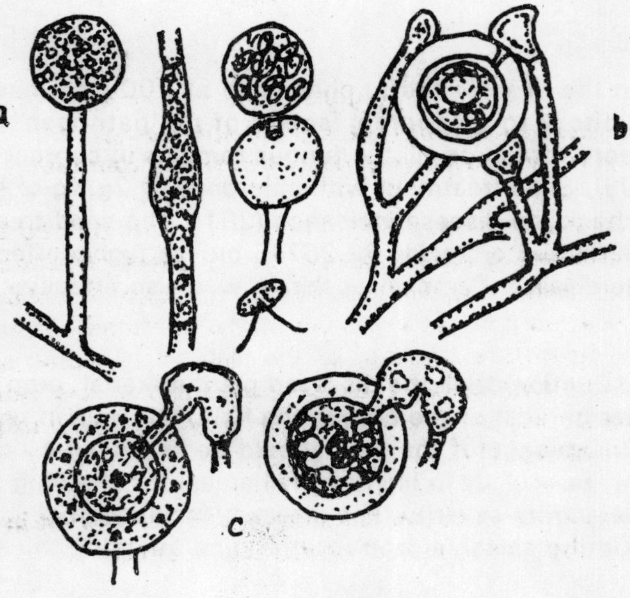

Fig.2.4
Pythium debaryanum : a. Terminal, intercalary and germinating sporangia and zoospore, b & c. Different stages of sexual reproduction.

The taxonomy of *Pythium* spp. was discussed in detail (315) and *Pythium splendens* var. *hawaiianum* was reported as parasitic on greengram.

Disease cycle and epidemiology :

The *Pythium* spp. infecting greengram are mainly soil-borne. Oospores persist in the soil resisting adverse conditions. Under sufficient moisture, they germinate and produce mycelium which later forms asexual stage i.e. sporangia. The mycelium is capable of infecting the host. The zoospores formed in the vesicles of sporangia are released in soil water to swim about and spread from place to place. Under adverse conditions for asexual multiplication, the sexual reproduction starts resulting in oospore formation.The activities of the pathogen at the seedling stage of host are favoured by warm wet overcast weather. Sporangia or oospores are splashed on to the plant from the soil and cause infection after germination. The soil collected around the infected plants is heavily infested. Growing blackgram plants in such soil results in pre- and post-emergence damping off to an extent of 76.1 and 11.6% respectively (132).

Control :

Chemical : In the *in vitro* tests, chloroneb at 200 ppm has been reported inhibitory to greengram isolate of the pathogen. It also reduces oospore formation till the fungus remains in contact of the fungicide (309). Seed treatment with chloroneb @ 2g/Kg of seed is reported to check the disease incidence (231). The seed treatment with CGA 48988 and orthocide RE 26745 proved highly effective in controlling the disease. Carboxin + thiram was also effective in this regard (82).

Cultural : Destruction of all the diseased plant material, crop debris and trash after harvest of the crop in the field, will help in reducing the inoculum. Drainage of the field should be improved by making suitable drains, partcularly in high rain fall areas, for checking spread of the disease and its severity. The disease incidence can be minimized by adopting suitable crop rotation and avoiding thick plant population.

2.2. SCLEROTIAL BLIGHT :

This disease was reported from Kanpur, India, causing considerable damage to greengram and blackgram crops (313). Both the crops were attacked under natural conditions. The sclerotial blight pathogen was later found attacking greengram by Roy also (287)

Symptoms :

The sympotms of the disease have been described at three stages as follows (313).

Pre-emergence stage : It includes rotting of ungerminated seeds and unemerged seedlings in the wet soil. The affected seeds become pulpy showing white fungal growth on their surface. Small, brownish, water-soaked patches appear on radicle and hypocotyle of seedlings before emergence. Later on, cotyledons are also involved and rot. The seedlings attacked at pre-emergence stage, never emerge out of the soil and are killed.

Post-emergence stage : The disease symptoms appear at the collar region near the ground level. Light brown, water-soaked discoloured patches apear on the place of attack. Later on, the diseased seedlings collapse and fall down as seen in case of damping off.

Mature plant stage : On the aged plant, circular or irregular, water-soaked spots appear first on lower leaves. These spots increase rapidly changing to dark brown or warm blackish brown. Such leaves show a wet rot and hang down. In severe cases, the wet rot may extend to petioles and tender twigs. Localized necrotic spots which are brown to dark brown in colour, appear on leaves in a comparatively drier weather. Hyphae of the fungus growing on the fallen leaves, form creamy to dark brown sclerotia at later stages.

Causal organism :

Sclerotinia sclerotiorum (Lib.) deBary

The young *mycelium* consists of sparsely branched hyphae which are hyaline, 2.6-4.2 μ in width and usually run parallel to one another. Older mycelium becomes rich in content, thick walled and closely septate measuring 3.1-5.2 μ in width. *Sclerotia* are tuberoid, globose and formed later on the rotten parts of the plant and also on the surface of the medium. They are cream coloured at first but acquire a darker shade when old. Sclerotia measure 1.0 to 3.0 mm in diameter. Though, the apothecia were not produced, the pathogen was identified as *S. sclerotiorum* (313) as depicted in Fig. 2.5.

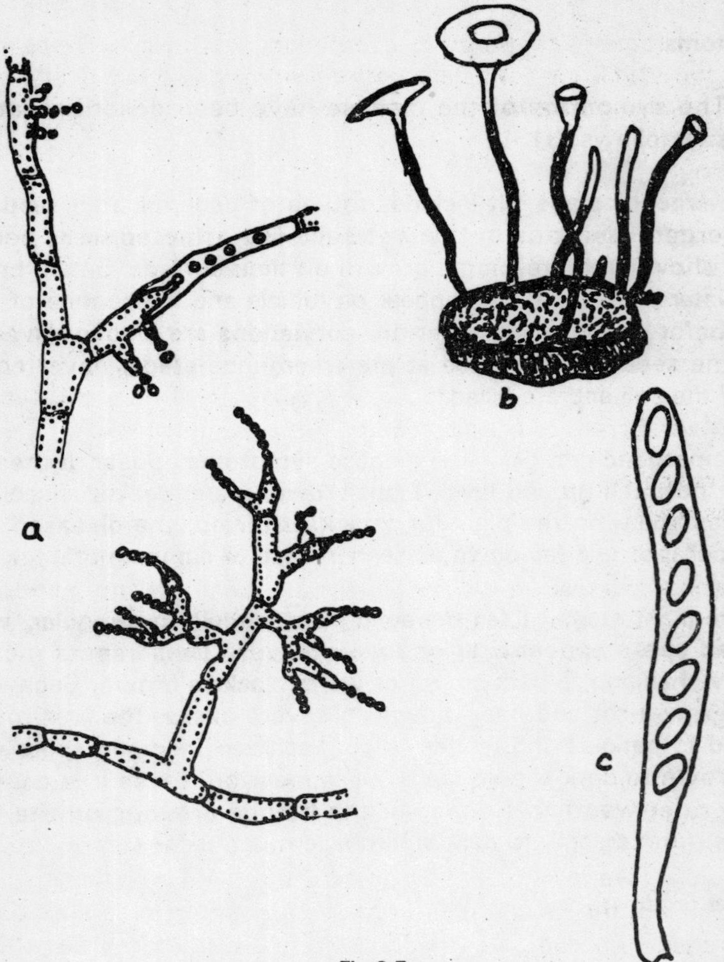

Fig.2.5
Sclerotinia sclerotiorum : a. Microconidia, b. Apothecia arising from sclerotium, c. Ascus. (all stages not observed).

Disease cycle and epidemiology :

The infection of the disease may spread from seed carrying the mycelium, sclerotia mixed with the seed, mycelial growth from sclerotia in the soil and frequently from the wind blown ascospores. The sclerotia are readily formed wherever dicotyledonous crops or weeds grow under humid conditions because of its wide host range and ability of surviving on decaying plant material. The sclerotia may remain viable in the upper soil layers for many years. The out break of the disease is favoured by crop injury due to climatic factors like high wind velocity etc. The pathogen has been reported to infect adzuki bean, chickpea and horsegram (231).

Control :

Chemical : Seed treatment with fungicides like thiram or captan may help in eliminating the seed-borne inoculum.

Cultural : Use of disease free seed, avoiding thick sowing, thorough weeding, destruction of the affected crop residues and crop rotation with cereals reduce the disease severity. Flooding the field with water is reported to kill the sclerotia after three weeks.

2.3. PHOMOPSIS SEEDLING BLIGHT :

This disease was first described from Annamalai nagar, India occurring severely on greengram and blackgram during February and March (268). The disease attacks mainly the seedlings at 2-6 leaf stages while older seedlings are not affected.

Symptoms :

First symptoms appear as water-soaked, greyish discolouration of leaf tips which spreads rapidly covering the whole leaf blade and the petiole. As the disease advances, the infected leaf droops. Later, it may involve the petioles, stem and the tender shoot finally killing the plant. In the field, patches of wilted plants may be seen distributed irregularly but in pots usually all the seedlings wilt within 48 hours after the first symptoms appear. On the petiole and stem, lesions are dark brown and elongated.

Causal organism :

Phomopsis sp.

The pathogen was found belonging to genus *Phomopsis* but differed from *Phomopsis vexans* occurring on brinjal (268). The *pycnidia* are formed on the infected leaves and are subepidermal, black, flask shaped, erumpent, ostiolate, 140-175 μ in size. The *pycnidiospores* are one celled, hyaline, thin walled, ellipsoidal or biguttulate, measuring 10.5 - 17.5 x 3.5-5.3 μ in size.

The fungus grows well on oat meal agar, PDA, nutrient agar and host extract agar media. In addition to greengram and blackgram, the pathogen is able to infect *Dolichos lablab and Raphinus sativus* causing localized leaf spots. It is unable to infect horsegram, chickpea, brinjal, tomato, chillies, ladies finger and mesta etc.

Disease cycle and epidemiology :

No work has been done but *Phomopsis* spp. have been found associated with seeds in many cases.

Control :

Not known

2.4. ACHAETOMIUM DISEASE :

Mortality of greengram seedlings was reported caused by a fungus to an extent of 58 per cent (373). The disease appeared in the form of pre-emergence and post-emergence seedling rot and blight after 5 and 15 days duration respectively.

Symptoms :

During the test, the commonest symptoms were pronounced with the occurrence of general browning on the internodes of severely attacked seedlings marked with stunting and distortion of coleoptiles (373). On removing the bran layer, the embryo showed dark brown, pin head strains in the initial condition which later coalesced to form patches of spots extending to hypocotyl and plumule at severity. They were invaded by actively growing hyphae and ascocarps resulting in embryonic axes to become jelley like mass indicating some diffusable toxins from the fungus to cause seedling deterioration.

Causal organism :

Achaetomium strumarium Rai *et al.*

Colony growth on the oat meal agar medium at $28 \pm 1°C$ is appressed, white at first becoming pinkish brown at later stages due to development of ascocarps. Reverse of the colony is yellow to brownish or blackish in colour. Aerial *mycelium* spreading, hyaline, septate, branched and 8-12 μ wide. *Ascocarps* develop after a week of incubation and measure 180-210 x 120-185 μ. They are dispersed or aggregated in groups, light grey to deep brown, flask shaped or ovoid, devoid of neck, superficial or semi-immersed, stromatic osteolate, osteolar neck absent with rhizoidal hyphae at the base. *Asci* clavate, fascicled, 8 spored, aporate, stipitate, evanescent, 50.5-56.8 x 6.5-10.8 μ. *Ascospores,* uniseriately arranged in ascus, greyish brown, elliptical, slightly epiculate at either end with unipolar pore, 8.4-16.5 x 6.2-9.9 μ in size. The germination of ascospores from one pole only.

Disease cycle and epidemiology :

The fungus is seed-borne in nature and attacks the seedlings of greengram at early stage only. No detailed information is available.

Control :

No control measures are recommended but seed treatment may help in eliminating the fungus from the seed.

2.5. FUSARIUM ROOT ROT AND WILT :

It is the minor disease of blackgram and greengram caused by species of *Fusarium*. The wilt of blackgram was first reported from Solan, India (350). Roy also reported *Fusarium* sp. occurring on blackgram (286). The root rot of greengram caused by *Fusarium solani* has been reported from Bangladesh also (30).

Symptoms :

Though, most of the *Fusarium* spp. cause seed rot reducing the germination of blackgram as well as greengram (424, 437, 442) some of them cause seedling mortality, root rot or wilt (300, 303, 350, 424).

The wilt of blackgram caused by *Fusarium* sp. has been described (350). In the initial stage, lower leaves become yellow gradually. Later, upper leaves are also discoloured followed by withering and severe shedding. In advanced stage, rotting and discolouration of secondary roots takes place.

The symptoms produced by *Fusarium semitectum* have been, described on blackgram (302). They include `damping off' of seedlings, drooping of tips and leaflets of young plants, leaflet chlorosis followed by wilting, drying and death of the plant. Initially, small, irregular, brown spots appear originating anywhere but generally starting from margins of leaflets. Spots spread covering major portion of the leaflet. The diseased parts dry up and become brittle while unaffected portion turns yellow and leaves dry producing typical wilting symptoms.

Causal organisms :

Fusarium semitectum Berk. & Rav.
Fusarium oxysporum Schl. ex Fries.
Fusarium solani (Mart.) App. & Wr.
Fusarium moniliforme Sheldon

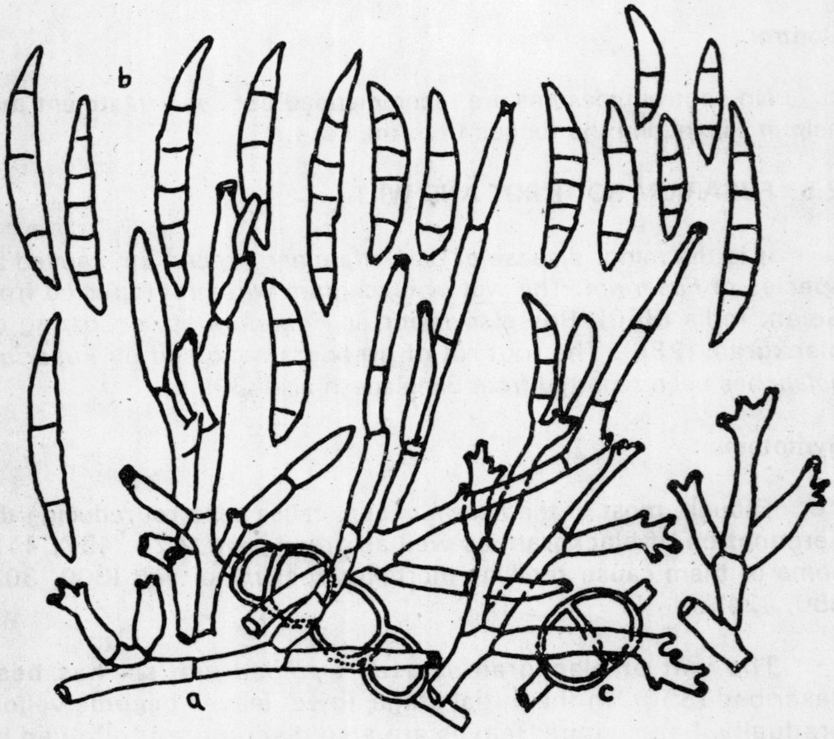

Fig.2.6
Fusarium semitectum : a. Conidiophores, b. Macroconidia, c. Chlamydospores.

F. semitectum (Fig. 2.6) and *F. oxysporum* (Fig. 2.7) were isolated from root, stem, leaflets and seeds of blackgram and greengram and were reported to cause wilt in both the crops in India (300). The last thee species have been reported to cause root rot and wilt of greengram (424). *F. semitectum* causes brown discolouration of xylem vessels of blackgram with presence of the mycelium in the vessels (302). The culture filtrate of *F. oxysporum* reduces the seed germination and inhibits root and shoot elongatin of blackgram (72). Roots are more sensitive than shoot for the inhibition effect. It has been found toxic to callus growth and weight in greengram variety DS 7.

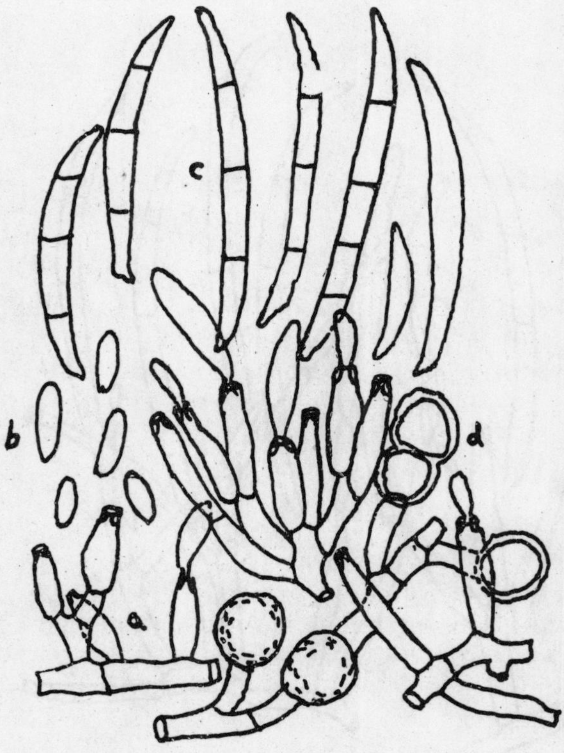

Fig:2.7.
Fusarium oxysporum : a. conidiophores, b. Microconidia, c. Macroconidia, d. Chlamydospores

However, *F. oxysporum* and *F. solani* are reported to be either weekly pathogenic on greengram (228) or not pathogenic at all (18). The extract of blackgram and greengram seedlings reduces the spore germination of *F. oxysporum* and *F. solani* (46). Hence, their pathogenicity needs further confirmation.

F. oxysporum reduces the number of nodules and their dry weight in blackgram (170). On the other hand, Bhargava *et al.* (46) reported that inoculation of greengram and blackgram checks the infection of roots by *Fusarium* spp. The plants with good inoculation were found free from infection by *Fusarium* spp. *F. moniliforme* (Fig.2.8) inhibits germination of greengram.

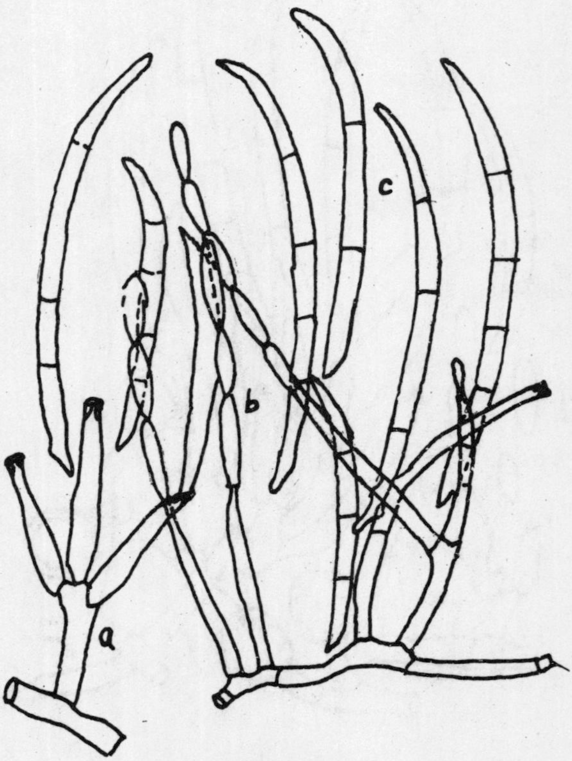

Fig.2.8.
Fusarium moniliforme : a. Conidiophores, b. Microconidia in chain, c. Macroconidia.

Control :

Chemical : Seed treatment with thiram or dithane M 45 @ 2-2.5 g per kg of seed is effective in eliminating the seed-borne *Fusarium* spp. of greengram (424, 425). Bavistin is highly effective *in vitro* but does not check the pathogen *in vivo.* The benomyl has been reported to be effective against *Fusarium* spp. However, the exudate of greengram reduces the toxicity of the benomyl against *F. solani* f. sp. *phaseoli* (428).

Biological : Some of *Aspergillus* spp. and *Trichoderma viride* have been reported to be antagonistic against *F. oxysporum. T. viride* checks the pre-emergence death of blackgram caused by *F.oxysporum* (229).

2.6. BLACK ROOT ROT :

The black root rot of greengram occurs particularly in cooler regions. The first report of the disease seems to be from Uzbekistan, U.S.S.R. (417). It has also been reported from Ontario, Canada (18). The disease is not common in most of the greengram growing areas.

Symptoms :

The symptoms are characterized by rotting of rootlets of greengram plants which initially develop purple lesions, turning black with age and leading to stunting of the host plant. Cortex and peridium tissues of roots are invaded. Entry of the pathogen is via ruptures of the cortex due to emergence of secondary root primordia. Distinct purple lesions are seen on tap root and some time on stem base.

Causal organism :

Thielaviopsis basicola (Berk & Br.) Ferr.

Fungus produces whitish layer of *conidiophores* under damp conditons. Conidiophore consists of short plump barrel shaped cells which terminate in a long tapering cell within which *phialoconidia* are borne endogenously and are 90-170 μ long and 5-6 μ wide at base and 3-5 μ at tip. The *condia* are hyaline, linear, 3-5 x 8-23 μ in size, becoming ovoid before germination. Conidia are known as microconidia. *Macroconidia (arthroconidia)* or *chlamydospores* are brown with rounded tips, formed in short chains from hyaline basal cells and are 14-16 μ in length (Fig. 2.9) At maturity they turn black and break apart (70).

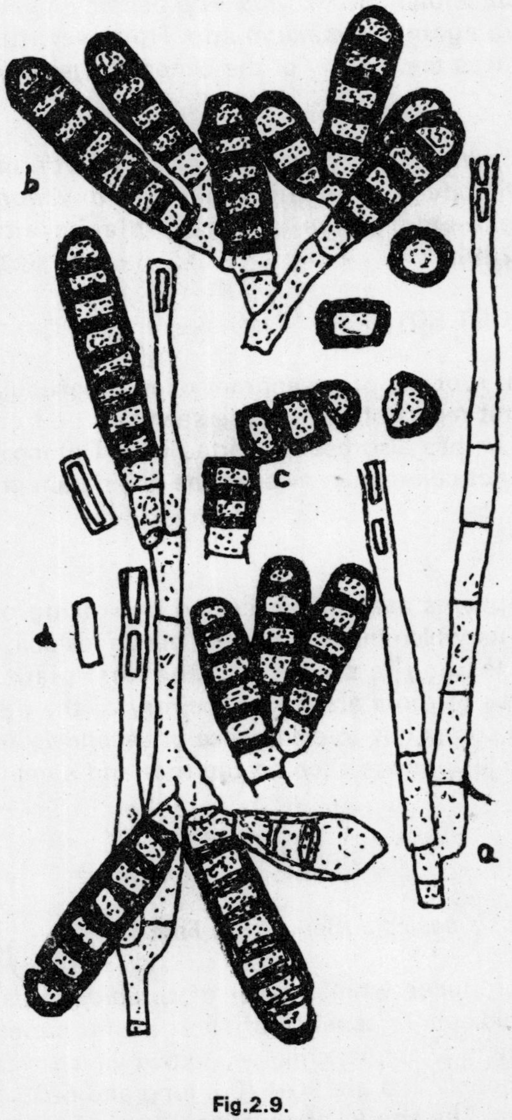

Fig.2.9.
Thielaviopsis basicola : a. Phialides, b. Arthroconidia, c. Separation of arthroconidia, d. Phialoconidia.

Disease cycle and epidemiology :

The organism is capable of prolonged saprophytic survival (70). The disease occurs during the early growing season and the wilting takes place during the flowering and late pod filling stages on clay and sandy soil in Canada (18). Temperature of 17-26°C, high soil moisture and in-adequate soil aeration favour the disease development.

Control :

Seed treatment with 50 % thiram and copper trichlorophenolate is reported to reduce the infection (417). The soil should be well drained and susceptible weed hosts should be removed.

In Korea, it has been reported that soybean isolate of *Calonectria crotalariae*,the causal agent of black root rot, is able to infect *Vigna radiata* also but no control measures are known (370).

2.7. BROWN STEM ROT :

This disease occurs on soybean mainly in U.S.A. where greengram is also attacked by it resulting in death of plants (57, 74).The disease is the main problem of adzuki bean in Japan where it damages greengram also (171).

Symptoms :

The fungus invades the vascular bundles in the roots and then reaches upto the stems and petioles turning the vascular bundles brown. The leaf tissues between the veins turn brown and dry rapidly. Plant starts wilting after flowering, dries and finally defoliates. Brown discolouration in side the root and lower stem is visible at early stage only on spliting and no external symptoms are seen.

Causal organism :

Cephalosporium gregatum (=Phialophora gregata)

On PDA, the *conidia* of soybean isolate from stem are 3.4-6.9 x 1.6-2.9 μ while those formed inside the infected host are 4.8-7.4 x

1.9-3.4 μ. They are single celled, hyaline, produced on tips and collected there in form of small droplets.

The pathogen produces gregatin A,B,C,D and E of which A,C and D are able to produce wilt, death of leaves and vascular browning of greengram cuttings (168). They do not cause vascular browning in soybean suggesting that soybean symptoms may be due to inhibition of photosynthetic electron transport by gregatin produced by the pathogen (281).

Disease cycle and epidemiology :

The work done on soybean indicates that the pathogen is soil inhabiting and causes losses under cooler conditions. However, *Cephalosporium* sp. has been found associated with blackgram seeds (427).

Control :

Not known in case of greengram and blackgram but in case of soybean, crop rotation is recommended in which host plant is to be grown after 3-4 years only (57).

2.8. STEM ROT :

In Japan, *Phytophthora vignae* is reported to cause the stem rot of *Phaseolus radiatus* var. *aurea (Vigna radiata)* and *Vigna sinensis.* Several isolates of this pathogen obtained from both the crops were compared for their pathogenicity. They were found to be highly specific in this regard (389). Hence, two different forma species are proposed of this pathogen. *Phytophthora vignae* f. sp. *adzukicola* Tsuchia *et al.* attacking *Vigna radiata* and *Phytophthora vignae* f. sp. *vignae* infecting cowpea. In addition of the differences in the host specificity, 3 races of *P. vignae* f. sp. *adzukicola* were recognized.

2.9. MACROPHOMINA DISEASE :

This disease is produced by a single pathogen in the form of charcoal rot, ashy stem blight, root rot,leaf blight and pod rot etc. and causes serious losses to greengram and blackgram crops. Root rot of greengram caused by *Rhizoctonia* sp. was found to occur in India as early as 1918 (55). Later, McRae (200) isolated the fungus *macrophomina phaseoli* from the wilted plants of greengram and blackgram in Madras, India. The disease occurs commonly in tropical and subtropical regions particularly in Uganda (109), United States (121), Sudan (375), Australia (23), Philippines (413), Pakistan (306) etc. In India, the disease is reported from all the greengram and blackgram growing states like Uttar Pradesh (199), Madhya Pradesh (247), Punjab (60), Orissa (437), Maharashtra (170), Karnataka (120), Haryana (98), Rajasthan (390), Tamil Nadu (399) etc.

Nature and extent of damage :

The disease appears in distinct phases attacking different plant parts at various stages of crop growth. Both the crops are damaged seriously due to seed rot, seedling blight at early stages and premature death and leaf blight at later stages. The stem infection of blackgram reduces number of pods per plant, 4.1-52.2% and 100 seed weight 3.5-11.4% (120). Plants infected at early stage show higher reduction in pod formation. Infection on foliage and pods reduces the yield as well as quality of seed. In Rajasthan, Macrophomina blight causes 24.07% loss in yield of greengram (390). In case of greengram, infection of seed may cause severe leaf blighting resulting in the yield loss up to 10.8% and reduction in total soluble sugars, reducing sugars and protein contents of the seed greatly (158). The infection of root by the pathogen causes severe reduction in the rhizobial nodulation and dry weight of plant (170).The pathogen causes deterioration of the outer layers of the root which may be responsible for the deterioration in the nitrogen fixing system of *Rhizobium* (182). It ultimately reduces the yield of the surviving plant drastically.

The extent of damage caused by the disease depends much on the time when it infects the plant. In Madhya Pradesh, this disease is reported to appear quite early in the season indicating that it gets sufficient time to damage the crop seriously (308).

Symptoms :

The disease appears in different phases on both the crops caused by the same pathogen (163, 308) and the symptoms can be described as below :

i **Charcoal rot** : The seedling-rot phase of this disease is commonly known as cahrcoal rot (231). After sowing, the seed or soil-borne inoculum of the pathogen causes seed decay and pre-and post-emergence rots resulting in damping off. Ungerminated seeds show rotting of radicles and plumules (248). Infection of the germinating seeds results in rupture of the seed coat exposing the cotyledons. Such seeds rot and disintegrate producing a foul smell. The post-emergence seedling rot is characterized by the rotting of the hypocotyle region, wilting and death of seedlings. Minute, circular, brown specks apear on cotyledonary leaves and leaflets, first on upper surface. The specks enlarge up to 1 mm in diameter, become grey in centre with a definite brown margins. Cankerous lesions may be formed on the hypocotyl region within 7 days after emergence (248). In case of the heavy infection of cotyledons and embryo, the seeds are rotted while the seedlings may show stunting of plumules and radicles if the infection is light (307).

ii. **Ashy stem blight** : Plants at later atages may get infection in the collar region forming greyish-black, sunken spots. The lesions may enlarge to form elongated cankers with grey centres and brown margins. They may girdle the stem which may break from the site of infection. Under certain conditions, the fungus from the collar region may spread fast upward in the stem and/or downward in the root resulting in death of the plant with ash coloured stem.

iii. **Stem canker** : These symptoms have been reported on blackgram 30 DAS under rice fallow conditions in Andhra Pradesh, India (421). Small, grey and suctiform lesions appear on stem and branches (Fig. 2.10). When fully developed, they measure 10 mm in length and later turn in to large, deep seated cankers with markedly verrucose surface. The cankerous growth is mostly confined to the collar region but also extends up to the internodal portion. Discolouration of tissues does not extend to roots. Profuse, white, cottony mycelial growth is seen on the internodal portion under dampy weather conditons. Some plants may be dwarfedl with dark green, leathery, puckered and some times curled leaves. Plants do not die immediately but flower and produce under-developed pods due to premature senescence. During the pathogenicity tests, only injured seedlings were infected producing typical symptoms.

iv. **Root rot** : Roots of the plant may be infected by the inoculum present under-ground or the infection may spread dawnward from the collar region. The root system of the infected plant shows

Fig.2.10.
Stem canker of black gram caused by *M. phaseolina*
(Courtesy : Dr. K. Abbaiah)

decaying of the tap and the secondary roots due to which it is easily pulled out from the soil. Root tips of the freshly wilted plant are slightly moist and sticky to touch. In adult plants, the cortex is shredded and wood is discoloured brown. In due course of time, the wilting plants are defoliated (248). Minute, black sclerotia may be seen on spliting or bark pealing of stem and root.

v. **Leaf spot/blight** : This phase of the disease was first reported from Madhya Pradesh, India, on greengram (247). It includes brown discolouration along the veins of the leaves starting with the cotyledonary leaves from the junction of the petiole and leaf lamina. Finally, the leaf becomes distorted and drops. Circular brown spots of 0.5-1.0 mm size appear on the upper surface of cotyledonary leaves and leaflets which later become greyish in centre with a definite brown margin. In the field, it appears on 4-6 week old plants (98). Usually, the older leaves are first attacked. Initially, small, circular to irregualr, brown to reddish brown lesions, 2-5 mm in diameter,appear on or near the margins which enlarge and coalesce. Under hot and humid conditions, the entire plant may blight (Fig.2.11). The dried leaves are yellowish brown and become papery. Severely affected leaves fall off prematurely without any fruiting body formation. In few cases, pycnidia may be detected on decomposing leaves lying on the soil. Stem or roots of the blighted plants are not affected though, petioles develop reddish brown streakes occasionally. Pod formation is greatly reduced.

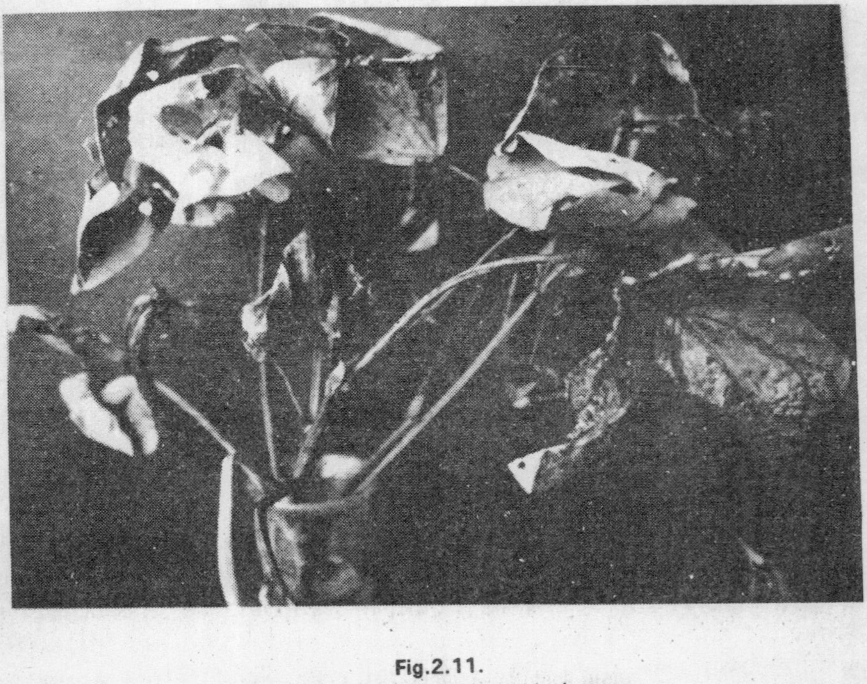

Fig.2.11.
Macrophomina leaf blight of black gram.

On blackgram, leaf spots were first observed in Madhya Pradesh, India (181). The disease appears on plants at 2-3 branch stage in the form of minute, brown spots which later enlarge up to 5mm in size, become irregular and striated with conspicuous reddish brown discolouration on both sides of the leaf. Such spots are found on greengram as well. In Tamil Nadu, the disease was observed on 10-20 days old plants (399) with symptoms resembling to those described on greengram (247). The veins of the top young leaves turn reddish brown (Fig. 2.12) on the upper surface. Later, the interveinal areas also turn reddish and lastly brownish. The infected leaves show characteristic twisting and distortion. The infected leaves slowly dry up and fall down. Subsequently, the petioles and stems are also affected. In severe cases, the whole plant is killed.

Fig.2.12.
Macrophomina phaseolina on mung bean : Infection of veins.

vi **Pod blight** : The pods may be infected becoming discoloured and shrivelled or blighted completely without seed formation. If the infection is after seed formation, discolouration of skin and shrivelling of grain is noted (Fig. 2.11). Seeds from infected plant carry the pathogen externally as well as internally which helps in survival of the pathogen during the off season (8).

Causal organism:

Macrophomina phaseolina (Tassi) Goid., the pycnidial form
Rhizoctonia bataticola (Taub.) Butler, the sclerotial form
Synonyms : *Macrophoma phaseolina* Tassi
Macrophoma phaseoli Mauble
Macrophomina phaseoli (Mauble) Ashby
Sclerotium bataticola Taub.

On wilted plants of greengram and blackgram, the pathogen was first recorded by McRae (200) who established the genetic connection between *Macrophomina phaseoli* and *Rhizoctonia bataticola*. Later, it was experimentally proved that *M.phaseoli* is the pycnidial state of *R.bataticola* on greengram (297).

Morphology : The fungus produces white fungal growth on media later changing to brown-black in the centre due to the formation of numerous small black sclerotia. The *mycelium* is hyaline to brown, branched, septate, dendroid and 6-8 μ thick (135). The *sclerotia* are formed on or within infected host tissues and are black, smooth, hard, 100-1000 μ in diameter (50-300 μ in culture). The sclerotia of blackgram isolate are 42-126 μ in size (135). The *pycnidia* are not formed in culture but are formed in large number on the host under certain set of conditions. They are dark brown, solitary or gregarious on leaves and stems, immersed, becoming erumpent, 100-200 μ in diameter with multicellular and pigmented wall and apical ostiole. *Conidiophores* (phialides) are hyaline, short, obpyriform to cylindrical, 5-13 X 4-6 μ in size. The *conidia* are hyaline, ellipsoid to obovoid, single celled, thin walled and 14-30 X 5-10 μ in size. Profused pycnidial development was noted on the greengram roots infected with the pathogen, when they were placed in moist chamber at the room temperature (248). They were black, more or less erumpent, globose or depressed, scattered, 150-200 μ in diameter, without stroma and with small osteole. The pycnidiospores were oval, or elliptical, hyaline, non septate, thin walled with one end often pointed and the other blunt, about 3 times as long as wide and 10-24 X 6-10 μ in size (Fig. 2.13). The pycnidiospores develop. *R.bataticola* in culture and germinate well at 25-30°C (247).

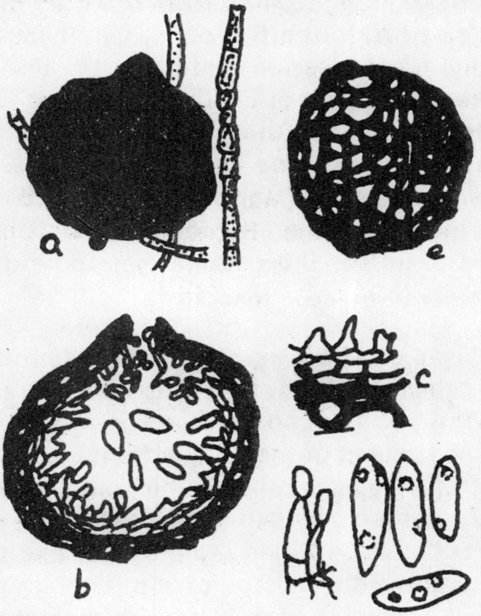

Fig.2.13.
Macrophomina phaseolina : a. Pycnidium b. Cross section of pycnidium, c. Conidiophores d. Spores and conidiophores bearing conidia. e. *Sclerotium*

The perfect state of the pathogen on greengram and blackgram has not yet been discovered. However, Ghosh *et al.* (89) described *Orbilia obscura* Ghosh *et al.*, a new species, as the perfect state of the jute isolate of *Rhizoctonia bataticola.*

Growth and pathogenicity : Asthana and Hawkers medium, 30°C temperature, 75 % RH and 5 pH are optimum for the growth. Maltose, lactose as the carbon source; KNO_2, KNO_3 and $NaNo_3$ as nitrogen source are best for the growth. Sclerotia formation is influenced by the depth of the medium, RH and the nitrogen source used (247). In culture medium, the growth of *R. bataticola* from greengram is supported by N,P,K but it has no correlation with the disease development. However, virulence is pre-conditioned by nutritional status of the substrate medium. Use of glucose and sucrose as the carbon source, asparagine as the nitrogen source and absence of bivalent metal ions (Fe^{++}, Mg^{++}, Zn^{++}) from the medium increases virulence of *R.bataticola* causing seedling mortality and foliage blight of mungbeam (123, 125). This effect on the pathogen does not influence the efficacy of carbendazim and thiophanate methyl fungicides for controlling the disease (124). The mungbean isolate grows best at 35°C temperature (98). The age of the inoculum also influences the virulence of the pathogen. The young inoculum of *M.phaseolina* in which sclerotial formation just starting, is more pathogenic than the old one (126). The pathogen produces certain metabolites harmful for the host. The culture filtrate has been proved toxic for the callus development of greengram variety Pusa Baisakhi (53).

Variability : Isolates of *M.phaseolina* from sesamum, horsegram, cotton, tobacco, chillies, citrus, potato, sunflower and cowpea are reported to infect greengram and/or blackgram (44, 61, 122, 297, 368, 369). But the isolates of the fungus from blackgram, groundnut and sesamum differ from each other in their pathogenicity (367). The blackgram isolate is more virulent than others. Similarly sesamum and castor isolate has been found more aggressive than groundnut isolate on greengram (98). Isolates of the pathogen from different plant species differ in respect of their growth characters and sclerotia formation also. The greengram isolate is quite different from the sorghum root isolate of the pathogen (245).

Six isolates of the pathogen from soil, seed, root, stem, leaf and pod of blackgram were reported to differ from each other in one or more of their characters of pathogenicity, morphology and growth requirements etc. (133,135,136,163). The soil isolate was the most pathogenic followed by root and pod isolate. The mycelial growth as well as the sclerotial size was least in the soil isolate. The stem isolate was the least pathogenic. It has been observed that the leaf isolate of the pathogen from greengram does not infect roots of the same host (98). In case of blackgram isolates also, the cross pathogenicity could not be proved on leaf and root of the same host (399). The sclerotia as well as pycnidia were large in case of leaf blight isolate (65-100 μ and 133-175 μ respectively) as compared to root rot isolate (55-80 μ and 120-138 μ respectively). Both of these isolates differed in their nutrient requirements and the enzyme production also (295). The root rot isolate preferred the host hypocotyl extract while the leaf blight isolate preferred leaf extract for their growth. Both the isolates produced pectolytic and cellulolytic enzymes *in vitro* and *in vivo* and addition of glucose increased the production of polygalacturonate trans-eliminase by root rot isolate while it was reduced in case of leaf blight isolate. Similarly, the stem and pod isolates from mungbean plant yielded less mycelial growth as compared to leaf, seed and root isolates while the sclerotial production was in reverse pattern (122). However, leaf isolates of the pathgen obtained from different localities did not differ from each other (142).

Host range: The fungus *R. bataticola* attacks about 284 plant species (454). In Madhya Pradesh, the greengram isolate of the pathogen has been found to infect blackgram, peas, cowpea, pigeonpea, okra, chillies, bean, and tomato (248). The leaf blight isolate from greengram caused seedling mortality in greengram, pigeonpea, blackgram, bean, faba bean, sunflower and *Ricinus communis*. The stem was infected in case of greengram, pigeonpea and *Vigna unguiculata* only. The leaf infection was noted in all the cases except *V.unguiculata* and *R.communis* (98).

Effect of host plant on the pathgen : The effect of root, stem and leaf extract of sunnhemp, blackgram, greengram, moth bean, *Sesbania* sp. and sesamum was assessed on the growth and sclerotia formation by the sesamum isolate of *M.phaseolina*. The blackgram stem extract did not allow the growth as well as the sclerotia formation, while the leaf extract inhibited the sclerotia formation with poor mycelial growth. The root extract of mungbean did not allow the sclerotia formation. Other host extracts allowed the growth and sclerotia formation to variable extent (169).

Host parasite interaction :

The fungus is reported to penetrate the root tissues of greengram within 24 hours. The fungal hyphae closely appress to the host surface along the longitudinal axis of the epidermal cell. After 72 hours, both, intra-and inter-cellular hyphae penetrate the cortical cells and the necrosis of the tissues is evident after 96-144 hours of infection. At the end of 168 hours, advanced type of necrosis is seen (248).

The pathogen produces pectolytic and cellulolytic enzymes which help in the pathogenesis (295). It also produces a toxin which interferes in the callus development of greengram (53).

After infection, the pathogen interferes with the metabolic activities of the plant (298). It is reported that greengram cultivars (PDM 116, MUG 125 and ML 337) with moderately resistant reaction against *M.phaseolina* contain low reducing sugars; high phenols, amino acids, chlorophyll `a' and `ab'. After infection by *M. phaseolina*, sugars, phenoles and amino acids increase but chlorophyll is reduced in all the cases. In susceptible variety K 851 and K 1, the sugars, phenoles and amino acids decrease.

In case of the infection of blackgram stem by *M.phaseolina* causing stem canker symptoms, the pathogen penetrates the host tissues after injury to host and not through the intact surface. The mycelium in such cases is detected intercellularly but in some cases it is found plugging the xylem vessels too (421).

Disease cycle and epidemiology :

The pathogen has been found associated externally as well as internally with the seeds of greengram and blackgram (8,9, 98, 133, 247, 422, 431 438). In greengram, 2.2-15.8% seed infection has been recorded but from diseased pods up to 61.8% infected seeds are obtained of which 47% may be internally infected (158). The infection of the blackgram seeds ranged from 12-58% in samples collected from Madhya Pradesh (9). In the infected seeds, the sclerotia and pycnidia of the pathogen may be observed either on the seed surface or under the seed coat (231). However, the pathogen has been detected from the seed coat, cotyledons, radicles and plumules of blackgram (9,307). The extent of infection is reported in decreasing order in different parts, from pod shell to embryo, indicating that the seed is infected after the localized infection on pod and not systemically (9). The pathogen can survive in the seed up to 36 months under laboratory conditions (158). The infected seeds and the pycnidia formed on root or defoliated leaves appear to be the main source of primary infection in the field. The pathogen is transmitted from seed to seedling. In addition to infected collar regions, the rotted seeds and detached infected cotyledons provide an extra source of the inoculum in the soil for root infection. The leaf blight and pod infection usually results from the secondary infection from the rain splash and air borne conidia or sclerotia (9,164). In the field, the pathogen can survive on infected plant debris in the soil. At room temperature the sclerotia are reported to survive for 54 months (367). The saprophytic survival is better in deep black or the medium type of soils (245).

The soil conditions influence the disease development greatly (248). It is favoured in sand+soil mixture and 40-60% soil moisture but reduced at 20% soil moisture in the mixture of sand + soil + organic matter. The pathogen grows vigorously in soil at 15% moisture and 30°C temperature but not at 25°C temperature and 20% soil moisture (152). Higher incidence of the disease has been observed (90%) in greengram grown in March when soil temperature ranges between 22-37°C as compared to 70% incidence in November with a soil temperature of 20-27°C (248). The disease is favoured by acidic condition (5 pH) and is reduced at 9 pH considerably.

Nutrition of plants influences the disease severity on greengram to a considerable extent (153). The infected seedlings die only 20% in 27 days if N,P,K, are supplied but the death may be 90% with in 10 days if no N,P,K or only P,K, is supplied. High dose of P_2O_5 is reported to increase the leaf blight severity also (342).

The quantity as well as the quality of the inoculum also play important role in the disease development (122). The increasing density of the inoculum increases the leaf blight intensity on greengram. Under artificial inoculation, 4-8 days old inoculum of *R.bataticola* is reported to cause severe foliage blight of mungbean at 30-35°C (98). The plant injury also favours the disease development greatly.

Thus, high temperature, high humidity for 8-12 hours and young inoculum are cardial for leaf blight development on greengram.

The leaf blight phase of the disease appears when the crop is 4-6 weeks old in presence of prolonged high humidity. Infact, the invasion of the mungbean plants by this pathogen is favoured by the devitalization characterstics of plants subjected to environmental extremes and wounds (121). Treatment of the seed with the colchicine 0.25% also increases the susceptibility of the seedlings of blackgram to *M.phaseolina* (255).

Host resistance :

Large number of germplasm types and cultivars have been evaluated for resistance against *M.phaseolina* in the field under natural disease occurrence and/or artificially inoculated conditions and the results are compiled in Table 2.1. The disease scoring has been done on the basis of percentage infection in case of seed/stem/root rots. The grading and grouping of different entries is mostly done as follows :

Grade	*Per cent disease incidence*
Free	= No disease
Resistant	= 0.1 to 5.0 % disease incidence
Moderately resistant	= 5.1 to 10.0 % disease incidence
Moderately susceptible	= 10.1 to 25.0 % disease incidence
Susceptible	= 25.1 to 50.0 % disease incidence
Highly susceptible	= 50.1 to 100 % disease incidence

Grading of entries for the leaf blight phase of the disease can be done on the basis of percentage leaf area involved (181) :

Grade	Description	Numerical value
O	Leaves disease free	0.0
I	Leaf area covered up to 5%	1.0
II	Leaf area covered 6-10%	1.5
III	Leaf area covered 11-25%	2.0
IV	Leaf area covered 26-50%	2.5
V	Leaf area covered 51-75%	3.0
VI	Leaf area covered 76-100%	4.0

The disease index (DI) can be calculated as

$$DI = \frac{\text{Sum of the numerical values of leaves X 100}}{\text{Number of leaves observed X 4.0}}$$

The varietal screening can be done easily by the 'Paper Towel Method' (65). In this method the paper towels are sterilized and they are impregnated with the suspension of the inoculum of *R. bataticola*.Surface sterilized seeds inoculated with the mycelial suspension are placed in the folds of the paper towels which are incubated at 25-28°C for 3 days after which observations on mortality are recorded. However, it has been reported that a variety found resistant to one phase of the disease may be highly susceptible to the other phase (399).

Table 2.1
Varietal resistance in greengram and blackgram against *Macrophomina phaseolina*

Crop	Entires tested	Disease phase	Entires found rsistant	Reference
Greengram	163	Seed rot and seedling blight	Nil	(66)
	Many	Root rot	LM 220 & MS 9385.	(400)
	14	Root/collar rot	PIMS 1, PIMS 2, PIMS 3, PIMS 4 and No. 11/395.	(419)
	24	Damping off/ root rot	12 entries were moderately resistant.	(52)
	21	Leaf blight	ML 10 & K 16-12-3.	(98)
	7	Seed rot/ seedling blight	NIL	(248)
	-	Leaf blight	BGG 1 (immune).	(311)
Blackgram	21	Leaf spot	T 27 & BR 68.	(181)
	91	Root rot	BR 68.	(325)
	17	Root rot	-	(401)
	Many	Root rot	Ac 258, No. 1792.	(399)
	Many	Leaf blight	US 131.	(399)
	102	Leaf blight	PLU 241, PLU 137, 1/1.	(104)

Control :

Chemical : Looking to the survival of *M. phaseolina* in seed, soil and its transmission through wind and rain splashes, large number of fungicides have been evaluated for use as seed treatment, soil drenching and spray on the crop.

Out of nine fungicides tested *in vitro* and *in vivo*, vitavax, demosan, benlate (0.25% each) and ceresan wet (0.1%) are found good for seed treatment of blackgram (134). In the *in vitro* test against *R. bataticola* from mungbean, carbendazim, benomyl, guazatine, dichlozoline, iprodione, and RH 893 are found highly toxic to mycelial growth (126), but the seedling mortality is checked by carbendazim, thiophanate-methyl, ceresan and captafol only when used as seed dressers. Thiophanate-methyl is found effective as soil drenching fungicide also.The topsin M (thiophanate-methyl) when used for seed treatment in greengram, moves apoplastically and maximum concentration is detected in cotyledons of 5 days old seedlings followed by aerial parts, hypocotyl and root (266). The concentration increases with age of the plant up to 15 days but no fungicide is detected in plant at harvest (266). Carbendazim is reported to be the best seed treating fungicide against *M. phaseolina* (8,173). In blackgram, the root rot phase of the disease is controlled by seed treatment with chloroneb which is reported to be absorved by the germinating seeds and accumulated in the roots of 12 days old seedlings (138). The damping off, root-and stem-rot caused by *M. phaseolina* is controlled by seed treatment with captan or thiram at 2.5 g a.i./Kg of seed and by the soil fumigation with 365 Kg a.i./ha of 57 % methyl bromide together with 43% chloropicrin (231). Brassicol (PCNB) and captan effectively reduce the saprophytic survival of *R. bataticola* in soil while bavistin and dithane M 45 are not effective in this respect (296).

The leaf blight phase of the disease could be controlled on mungbean by 2 sprays of benlate, captan, dithane Z 78 or dithane M 45 given at 10 and 20 days after sowing which maintained more than 90% healthy seedling population (51). However, most of the workers reported carbendazim (bavistin) as the best spray fungicide controlling the leaf blight phase of the disease and increasing the yield economically (124, 126, 172, 173, 341, 383, 384, 418). Though costly, it takes care of powdery mildew also and being systemic in nature it suits to the rainy season crops like greengram and blackgram (341). Hence, it has been recommended to treat the seed by carbendazim (1 g/Kg seed) or thiram (2.5 g/Kg seed) followed by 2 sprays of carbendazim (0.02-0.05%) on 30 and 45 days after sowing (172, 173, 341, 383, 418). It reduces the damping off, leaf blight as well as pod blight phases of the disease (8). However, it has been suggested that benzimidazole fungicides having similar toxophores, act differently for the disease control in different host and parasite combinations (374).

Cultural : Since the pathogen has very wide host range, the control of the disease is very difficult and requires the integrated approach. In addition to the chemical control measures, adoption of the following methods will help in reducing the disease greatly :

1. Avoidance of extreme soil moisture by providing good drainage of water.

2. Correct time of planting and avoiding the thick planting.
3. Adoption of crop rotation with resistant crop such as field pea, rice, sorghum, pearl millet etc.
4. Adoption of clean cultivation by destroying the diseased crop residues and proper weeding.
5. Adoption of cultural practices for increasing the crop health and vigour. Liming the soil and increasing the soil organic matter helps in controlling the disease (121).
6. Use of balanced fertilizer. Supply of N,P,K reduces the seedling death while their defficiency increases the disease incidence greatly (153).

Biological : Several microorganisms have been reported antagonistic to *R. bataticola* (68). Out of 38 fungi isolated from the rhizoplane, rhizosphere and non rhizosphere soil of blackgram,*Aspergillus aculeatus, Cephalosporium humicola, Trichoderma lignorum* have been found antagonistic to *R. bataticola in vitro. Arachniotus* sp, or *A. aculeatus* when applied to seed, controls the pre-and post-emergence damping off of blackgram due to *R. bataticola*. Both the antagonists control the pathogen when externally seed-borne but not internally. It is also influenced by the environmental conditions and age of the plant.

Epicoccum purpurascens has been reported to be a strong antagonist against *M. phaseolina* causing charcoal rot in greengram (336). It inhibits the disease development in unsterilized soil where it can suppress the pathogen up to 5 months. The culture filtrate of the antagonist is reported to inhibit the spore germination completely, to reduce the germ tube elongation, to inhibit the mycelial growth and to delay the sclerotia formation. It also inhibits the infection of mungbean seedlings as well as adult plants by *M. phaseolina* (333). Out of 18 isolates of the antagonist classified in to 4 groups, only few belonging to group II gave maximum protection to greengram plants against *M. phaseolina* on different parts of the plant (334, 335).

Trichoderma viride, when applied to the seed is able to check pre-emergence death due to *M. phaseolina* (229). *T. viride* and *Trichoderma harzianum* multiplied on the groundnut shell when added to the field @150 Kg/ha, reduce the root-rot due to *M. phaseolina* in greengram and blackgram (87, 294).

Similarly in the green house test, the indigenous isolate of *Rhizobium* from pea, lucern and soybean, reduced severity of the *Macrophomina* root rot in mungbean (415).

2.10. RHIZOCTONIA DISEASE :

It is a wide spread disease of pulse crops including greengram and blackgram causing symptoms at various growth stages of crop. Three main phases of the disease have been reported i.e. seed rot; seedling mortality due to stem canker/crown rot; web blight. All of these symptoms may not appear under a particular set of conditions. The Rhizoctonia blight on greengram seems to be reported first from Philippines (217) while the first record of the blackgram infection is from Maharashtra, India (392). One or more phases of the disease have been reported occurring in United State (128), Canada (18), Iran (145), Malaysia (3) Pakistan (13), Taiwan (188), Uzbekistan (417) and several other countries of the world. In India, the disease causes considerable damage to both the crops producing all phases of the disease. The disease was observed to reduce 4-57% population of different varieties of greengram in Iran (143) while in India 17-90% infection has been reported in both the crops (267, 288).The web blight phase of the disease has become the most serious problem of these crops in north India causing extensive damage (291).

Symptoms :

The symptoms can be described under three distinct phases of the disease, all of which may not appear under particular set of conditions :

i. **Seed rot** : These are very common symptoms observed in the infested soil. It has been reported that all the ungerminated seeds of greengram, when sown in the soil infested with the pathogen, are killed before their emergence (143). After germination the death is less.

ii. **Seedling mortality** : Two to three weeks old seedlings are infected mostly at the collar region producing stem canker symptoms and ultimately resulting in death. The stem canker symptoms on greengram have been described by Kaiser (143) and include appearance of small, reddish brown lesions in cortex of hypocotyl at or below the soil level. The lesions (cankers) enlarge with the age of seedling and become dark brown. The seedling is killed when the hypocotyl lesions girdle the stem after their coalesce or the enlargement. Prior to collapse of the seedling, the foliage becomes yellowish green and the leaves droop. The roots are generally not discoloured and appear healthy. Some times the infection at the collar region may also extend downwards affecting roots and resulting in death of the plant (231). From the necrotic roots, certain other fungi (*Macrophomina phaseolina, Pythium* spp. and *Fusarium* spp.) have been isolaed (429).

According to Sharma *et al.* (310), the seedling blight symptoms on greengram are of 2 types. The first category includes the symptoms on cotyledons and plumules. The infected seedlings emerge later than the healthy ones. The soil particles stick to the seed with rusty lesions on the cotyledons due to which they shrivel. Rusty lesions also occur on first true leaves. In case of heavy infection, the growing point is killed. The second category of symptoms appear in the form of rusty lesions of varying size on the stem. If the lesions girdle the stem completely, the seedling breaks easily at the point of infection. If the lesions are restricted, little damage is done to the plant. In case of blackgram, the brown coloured lesions appear on the hypocotyl and the discolouration extends to petiole and midrib (267).

iii. **Web blight** : The web blight symptoms have been reported on the foliage of greengram (81, 404) as well as blackgram (293). The symptoms start with the development of small, irregular, water-soaked, greenish pale spots with damp appearance on any portion of leaf and the stem. The spots turn dark brown, leaflets shrivel and dry up (Fig.2.14). During the period of high humidity, the disease spreads very fast. In the begining, the infection occurs on the basal parts including petiole of the lower leaf from where it extends towards the apex. The spots increase in size very rapidly covering greater parts of the leaf blad and stem which is a very conspicuous and destructive phase of the disease. The coloured, spider web like mycelial growth and white microsclerotia develop in abundance on the affected parts. The sclerotia later turn brown to chestnut brown within 2-3 days. All the above ground parts including pods are attacked (288). The infected branches shrivel at the point of infection, their leaves loose the normal green colour and wither. Affected parts rot rapidly in wet weather. Leaves and branches fall down leaving behind the main

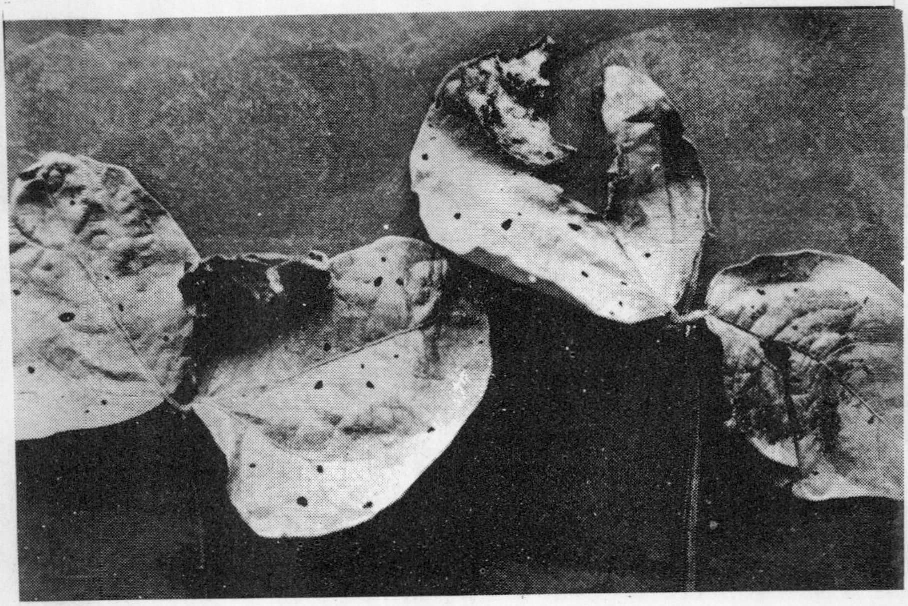

Fig.2.14.
Leaf Blight (web blight) of *greengram* caused by *Thanatephorus cucumeris*.

stem bare and in dry weather they may remain attached to the stem for few days. Infected pods are shrivelled and turn brown. If the pods are infected in the initial stage, the seeds are not formed at all while in case of late infection the seeds formed are very much shrivelled and underdeveloped. In severe cases the plants are killed before the flowering.

Causal organism :

Thanatephorus cucumeris (Frank) Donk.
Imperfect state : *Rhizoctonia solani* Kuhn.
Synonyms: *Hypochnus cucumeris* Frank
Corticium solani (Prill & Delacr.) Pilat.
C. sasakii (Shirai) Matsumoto
Pellicularia filamentosa (Pat.) Rogers
Rhizoctonia microsclerotia Matz.
and several others

The pathogen has been reported on greengram as well as blackgram under the name *T.cucumeris* (81, 293, 404) and *Corticium sasakii* (288, 290).

The colonies of *Rhizoctonia solani* are white and effused and later becoming light buff to brown. Sectoring of colony is frequent on agar media and the sclerotia may or may not be formed. The *mycelium* is fast growing, septate, hyaline when young but later turning brown, 4.5-8.5 μ in width. The lateral branches are constricted at their point of origin and possess septum near its junction with the main axis. In older hyphae, the branching is at wider angle but in younger one it is at the acute angle. The cells of growing hyphae are always multinucleate and the septum is dolipore type. *Sclerotia* are white but later turning chestnut brown, globose to subglobose, oval or cushion shaped, 2-3.5 mm in size but some times aggregating together to form masses up to 1.2 cm in diameter and are formed superficially (288, 290). The sclerotia are not differentiated in to rind and medulla. The *chlamydospores* are 12-21 x 9-13 μ in size and are formed in chains (96).

On the host, the sclerotia are superficial, variable in shape and size, black, scab like and accompanied with dark short celled, abundantly branched, stout mycelium without clamp connections. Inside the host tissues, the mycelium is slender, hyaline and with longer cells.

The perfect state *T. cucumeris* (Fig. 2.15) develops on healthy tissues adjacent to the lesion on blackgram leaves as well as *in vitro* (293).It is induced on the host under warm and humid conditions with heavy dew formation in the night. Fructifications are resupinate, creamy to greyish-white, loosely attached to the substratum, composed of archnoid, repent hyphae giving rise to thin hypochnoid or submembranous fertile patches. Clusters of basidia are produced terminally in a discontinuous *hymenium*. Hyphae 4.5-17 μ wide, without clamp connections, often anastomosed; basal hyphae widest with branching at wide angles while median and subhyaline hyphae are narrow, thin walled and hyaline. *Basidia* are mostly 9-25 x 5-12 μ in size but some times short,widened, barrel shaped to subcylindrical without a median constriction. *Sterigmata* are straight, attenuate towards the apex usually 4 per basidium, rarely 2,3 or up to 7 in number. *Basidiospores* are hyaline, smooth thin-walled, oblong to ellipsoid and dorsally flattened or broad ovoid and commonly widest toward the distal end and form secondary basidiospores (449).

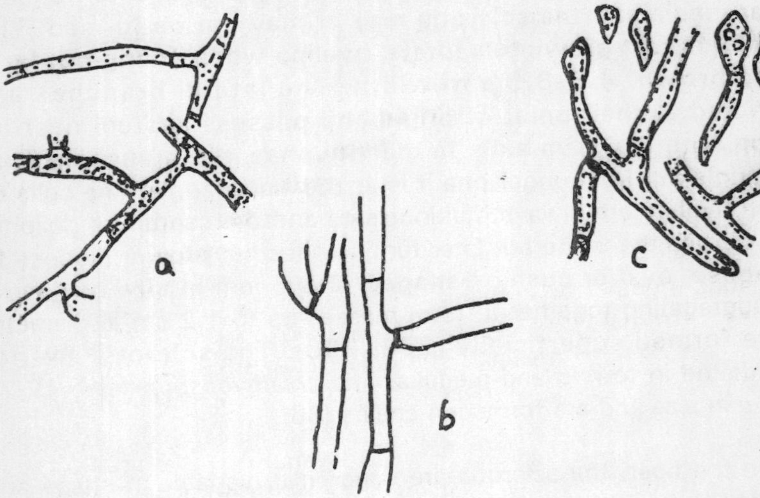

Fig.2.15.
Thanatephorus cucumeris : a, b. Cells of mycelium, c. Basidial formation, basidiospores and their germination.

Variability : The pathogen has very wide host range belonging to different families. Based on their cultural characters and pathogenicity, Housten (128) identified 52 isolates of *Corticium solani* from 260 collections for further pathogenicity tests on plant species including blackgram. In France, 38 isolates of *C. solani* could be grouped in to 5 on the basis of temperature requirements and virulence on various hosts including greengram (39). The blackgram isolate of *R. solani* is reported to differ from the lablab isolate (267). Parmeter (233) separated four groups of isolates of *R. solani* on the basis of their anastomosis ability and pathological behaviour. Ordinarily, isolates from the graminaceous hosts do not infect the members of Leguminosae but maize isolate is reported to infect several including greengram (34). The isolates from different legumes however, do not differ in their pathogenicity on greengram (146). Attempts have been made to find the differences between the isolates of *R. solani* obtained from different parts of the same host. Root and leaf isolates from greengram cause severe pre-and post-emergence damping off and pod rot but the seed isolate is less virulent (414). However, isolates of *R. solani* causing web blight in greengram, blackgram, pigeonpea and several other hosts do not show any remarkable difference (81.291).

Different field isolates of *R. solani* showing varied cultural characters and pathogenicity including tissue specificity have been considered as physiologic races of the pathogen.

Host parasite interaction :

It was thought that *R. solani* penetrates the host through stomata by means of growing hyphal tips without formation of any swelling, mycelial cushion or appressorium but now it has been shown (140) that the pathogenic isolate of *R. solani* forms the infection cushion on greengram hypocotyl. Non pathogenic isolates do not form any infection cushion. Not only the host, the infection cushion formation differs on different parts of the same plant according to their susceptibility. It has been reported that in both the crops, *R. solani* penetrates by hyphal tips developed beneath the infection cushion formed by the mycelial growth on the host surface and through the stomatal openings without the formation of appressorium above the stomata (71). After penetration, the hyphal branches grow rapidly both inter-and intra-cellularly in the host tissues. The invaded mesophyll cells collapse and become discoloured after some time.

After infection, the physiology of the host is influenced. It is reported to increase the rate of leakage of electrolyte from the cells of greengram hypocotyl. It is due to increase in the cell membrane permeability caused by a heat-labile substance detected in droplets formed on the infected host surface (184).It seems that the toxins produced by the pathogen play important role in the establishment of the infection of *R. solani* in greengram, Wu (408) showed a direct correlation between the growth, pathogenicity and toxin production in an isolate of *R. solani*.The gibberellic acid probably reverses the process of the disease development in greengram due to *R. solani*.

The infection by *R.solani* increases the reducing sugars and amino acid contents but decreases the nucleic acid and protein contents of greengram (409). The protein content decreases in the early stages of infection in cotyledons but increases in axil part with the development of the lesion (410). At this stage, flow of arginine and methionine from cotyledons to axil part has been found to increase. It is due to the fact that the pathogen is able to synthesize at least 15 amino acids of which DL-threonine stimulates the growth of greengram while DL-alanine, L-serine, L-cystine and DL-methionine are inhibitory (411).

Exposure of seed to *R.solani* increases the activity of phosphorylase and glutamate dehydrogenase but later it comes to the level of control (412). The aldolase activity is declined while glucose 6-phosphate dehydrogenase decreases first but increases later until symptoms appear. Amylase activity is increased. It is suggested that in the early stages of the pathogenesis, an increase in reducing sugars is caused by reduction in carbohydrate metabolism while later the hexose monophosphate pathway plays a major role in carbohydrate metabolism.

The total phenols content was reported more in healthy tissues of greengram as compared to those inoculated by *R.solani* (26). However, it resulted in apearance of two new substances while one disappeared after 24 hours of inoculation. The total phenolic contents of the infected hypocotyl of the greengram can be increased by treating it with 50 ppm ethrol (25). In the infected hypocotyl, the peroxidase activity increases at 24 hours after inoculation, decreases at 48 hours and 72 hours and again increases at 120 hours after inoculation. Polyphenoloxidase activity increases after inoculation and is correlated with the appearance of the symptoms (28). Treatment of the greengram seedlings/hypocotyl with ethephon, an ethylene releasing compound, is reported to enhance the peroxidase and polyphenoloxidase activity increasing resistance against the pathogen.

Disease cycle and epidemiology :

Though, the pathogen *R.solani* has been isolated from seeds of both the crops, it is primarily soil-borne and the initial inoculum comes mainly from soil splashed on to the leaves during heavy rains. The pathogen is capable of extensive growth through soil and survives on crop debris. It can persist in different soils for many years (231). The highest inoculum potential is noted in the top 10 cm soil and no inoculum is found below 40 cm (146). The fungus can grow in the soil very fast to a longer distance and can survive there in mycelial or sclerotial form (291). Once the floral parts are infected, the disease becomes seed borne which can lead to spread of the pathogen to uninfested areas. Camporata (56) evolved a technique to measure the infectious potential of soils infested by *R.solani* and calculated the linear regression of disease index against log soil concentration.

The collateral weed hosts play an important role in initiation and early spread of the disease to the main host (292). Because of their proximity to the soil carrying primary inoculum and their special microclimate, the weed hosts are first to take the infection and facilitate the production of the basidiospores of the pathogen. There after the pathogen becomes air borne to cause leaf, stem or pod infection and produces fresh crops of basidiospores on the main host. On blackgram, maximum basidiospore production and their discharge occurs during mid night and early morning before sunrise (293). The spore production is favoured by night temperature below 24°C, RH above 95% and rate of evaporation below 1.5 mm per day. In nature, basidiospore production starts in the mid August in north India. The basidiospores germinate in 2 hours and penetrate leaf through the intact surface by formation of infection cushion or by direct penetration through stomatal openings. They loose their viability rapidly in dry storage and after exposure to direct sunlight.

The disease develops fast at mean temperature of 20.5-30.5°C and 86-100% RH. The most suited temperature is 26-28°C and RH 100 per cent (292). The rapid spread of the disease on greengram has been noted under humid weather with a temperature range of 25-32°C.

The secondary spread of the disease may be due to the growth of filamentous hyphae from leaf to leaf and then from plant to plant in contact when the crop canopy is dense (291).

The susceptibility of greengram seedlings to damping off caused by *R.solani* decreases with the age (146). The root rot is also reported to be prevalent during the early growing season of greengram particularly in clay soil (18). However, for the web blight phase of the disease, the plants are susceptible from the seedling stage until maturity and the severity is maximum in 30-70 days old plants. It is mainly due to the dense crop canopy formed at this stage which facilitate the easy spread of the disease (291).

Varietal resistance :

In Iran, 40 genotypes of greengram were tested against *R.solani* under glass house conditions but none was found resistant (143,146). In India also no resistance could be identified in greengram varieties which showed 85-90% disease incidence with

25-30% mortality due to *R.solani* (288). However, sources of resistance have been identified in greengram at AVRDC, Taiwan (231). It has been reported that varieties ML 26, ML 9, 11157 and ML 70-10 are resistant to *Cercospora cruenta, Xanthomonas phaseoli,* MYMV and *R.solani* (137).

Control :

Chemical : It has been reported that in the area infested with *R.solani*, the plant stand is increased when the seed is treated with dexon-terrachlor (82% stand) and terrachlor (79% stand) as compared to 14% stand in control after 9 weeks of sowing (145).

Carboxin (vitavax) is reported highly effective against *R.solani* on blackgram when used as soil and /or seed treatment (38,97). It has been reported to be the second best fungicide as seed dresser for the control of damping off of greengram caused by *R.solani* but the seedlings are not protected against the collar infection (310). On the other hand it has also been reported that the vitavax checks the mycelial growth of *R.solani in vitro* but it is not much effective as seed dresser *in vivo* (139, 156). The change in polypheno constituents has been observed in the seedlings of greengram when they are treated with vitavax after infection with *R.solani*. These phenols are suggested to play an important role in protecting the seedlings from the pathogen (27).

Benomyl is also reported to give 100% protection to greengram seedlings when treated seed is sown in the soil inoculated with *R.solani*(139). Out of the emerged seedlings, 93.4% survived up to 3 weeks. Fungicides belonging to the benzimidazole group, particularly benomyl, carbendazim and thiabendazole, have been found very much effective seed dresseres/soil drenchers against *R.solani* on both the crops (156, 193, 289, 310, 374). The treatment of greengram seed with carbendazim @ 1 g/kg of seed, controls *R. solani* best in the moist soil (108). The antifungal activity of carbendazim is better at pH 7-8, temperature 20°C and sandy loam soil as compared to clay loam soil. When used as seed dresser, the benomyl first accumulates in the seed-coat in short time and then in cotyledons. Later, it accumulates in roots, stems and leaves with maximum concentration in roots (192). The ED 50 value of the benomyl against *R. solani* is 0.57 µg/ml (193)_. It is reported to inhibit formation of the infection cushions by *R.solani* on greengram hypocotyl when seedlings are dipped in 250 μM suspension for 30 minutes (154). Unchanged benomyl and its breakdown products have been detected in greengram seedlings after soil treatment with benomyl, however, hydrolysis of benomyl to its toxic derivatives in plant tissues is not rapid (141).

Amongst the non-systemic fungicides, brassicol (PCNB) is reported effective against *R.solani* on both the crops when used as seed dressing or soil drenching chemical (139, 154, 414). It checks the formation of infection cushions of the pathogen on hypocotyl of greengram and blackgram when seedlings are dipped in 500 μM suspension for 4 hours (154). Elazegmi and Mew (82) reported PCNB + thiabendazole as the most promising seed dressers against *R.solani* on greengram, soybean and cowpea.

Demosan and thiophanate methyl have also been found effective seed dressers (154, 156). The chloroneb is absorbed by the blackgram seed and moves systemically to growing parts. It accumulates maximum in roots and controls root rot caused by *R.solani* (138). Seed treatment with thiram 50% and coppertrichlorophenolate also reduces the infection by *R.solani* in greengram (417).

The vapours of aqueous extract of rhizomes of *Alpina carinata* are highly effective against the pathogen. The oil isolated from the rhizomes is reported to be effective even at 3000 ppm without any phytotoxicity to greengram seeds (167). The seed treatment with the mineral oils, EWOS and E 9267 controls more than 90% pre- and post-emergence damping off in greengram (155). The mustard oil is also effective but controls only the pre-emergence damping off.

The web blight phase of the disease could be contolled to an extent of 96% by spraying the benomyl or carbendazim 0.05% (289). The first spray was given 48 hours after inoculation and the second, 10 days after the first one. In Malaysia, benomyl reduced the spread of the disease when sprayed at the first appearance (3).

However, the fungus can adapt to a fungicide and hence, modified cultural practices in combination with other control methods may be helpful (231).

Cultural practices : Certain practices like proper sanitation, planting at a time to avoid rainy season during the susceptible crop stage, avoiding the thick canopy of crop by using proper seed rate and adopting proper crop rotation help in controlling the disease to a greater extent (217). The greengram seedlings are infected least in soil amended with the wheat straw (204). Burying the infected leaves immediately after the harvest will reduce the primary infection. For proper control of disease, the diseased material should be destroyed and crop rotation be practiced. Weed hosts should be removed, soil drainage should be good and soil should be well aerated. Shallow planting will also help in minimizing the infection by *R. solani* (231).

Biological : It has been found that *Trichoderma viride* Pers, ex Fr., a soil inhabiting fungus can destroy the pathogen and act as a biological control agent.

2.11. SCLEROTIUM DISEASE :

This is also a multiphase disease in which seed, root, stem as well as leaves are affected. The pathogen has a very wide host range attacking plants mostly at the seedling stage causing seedling blight or infecting the collar region of the plant. Infection of greengram by this pathogen was first reported from Japan (84). Later, the disease was reported from Trinidad under natural conditions (54). On blackgram, the disease was observed in India (278). The pathogen may infect the foliage of plant (206) during the rainy season. The disease is wide spread in the moist tropics and the warm temperate regions particularly in most of the Africa, Southern Asia, Australia, Central America, much of South America, U.S.A. and West Indies (CMI distribution map No. 311).

Symptoms:

The symptoms of the disease are produced at three distinct stages all of which may not be observed in a crop.

i. **Seed rot and seedling blight** : The seed may rot before emergence from the soil showing white fungal growth. The seedling blight symptoms are very common in which the infection begins at soil level causing damping off of infected seedlings.

ii. **Stem/Collar rot** : In adult plants, the infection occurs at the collar region from where it extends few centimeters above and below resulting in rotting of stem and roots (Fig. 2.16). The affected parts of the stem and roots rot at the soil level with dark brown lesions appearing adjacent or below soil level. The lesions may girdle the stem and later become covered with white mycelial growth in which white-brown mustard seed like sclerotia are embeded. The young infected plants die quickly after attack but older plants turn yellow. In severe cases, the leaves are discoloured and the infected plants wilt and die (129).

Fig.2.16.
Stem rot of green gram caused by *Sclerotium rolfsii*.
(Courtesy : Dr. K. Abbaiah)

iii. **Leaf spots** : These symptoms were first described from Madhya Pradesh, India and have been reported from Malaysia also (206 & 3 respectively). The symptoms include development of large necrotic leaf spots up to 3 cm in diameter. They are marked with the concentric rings in dead tissues which are more distinct on the upper surface. When young, the spots are small, greenish yellow, later becoming brown and enlarged. The smaller spots may coalesce to cover large area of the leaf. Whitish mycelium is prevalent on the lower surface of leaves radiating from the centre of spot towards the margins beyond the necrotic zone covering healthy area of the leaf. The central portion of the spot later fall away showing shot hole symptoms. Under heavy infection, the infected leaves dry and defoliate.

Causal organism :

Sclerotium rolfsii Sacc.
Perfect state : *Corticium rolfsii* (Sacc.) Curzi.
Synonym : *Hypochnus centrifugus* (Lev.) Tul. (84)

On PDA, the fungus produces fan shaped, profuse mycelial growth. The *mycelium* is hyaline, branched, septate, slender, with clamp connections. It has the tendency to grow on the glass surface and to aggregate in to rhizomorph cords. Clamp connections may be lacking in slender portion of the hyphae. *Sclerotia* are at first white and felty but later becoming brown, spherical and resemble with the mustard seeds. They are easily detached from the mycelium and measure 0.5 to 0.8 mm in size (Fig. 2.17). The perfect stage of the fungus is reported to be *Corticium rolfsii* (32).

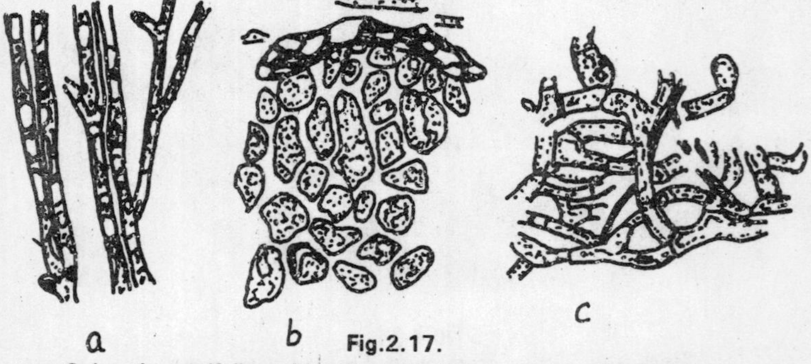

Fig.2.17.
Sclerotium rolfsii : a. Mycelium, b. Cross section of Sclerotium c. Cross section of germinating sclerotium.

Host parasite interaction :

The pathogen produces the enzyme, phosphatidase which inhibits the oxygen uptake in greengram causing damage to the plant (387). It also interferes with the rhizobial activity reducing number of root nodules and the dry weight of the plant but this reduction is less if *Rhizobium* is also inoculated (170).

Disease cycle and epidemiology :

The pathogen survives mainly on infected crop residues in the field and the weed hosts which are the sources of the primary inoculum. The survival of the pathogen is facilitated by the presence of undecomposed organic matter in the soil. The disease causes much harm in the light, sandy and wet soil. Shading and crowding increases the disease incidence further. The sclerotia of the fungus are dispersed through cultural practices, wind and water. They may survive passage through sheep and cattle. The disease is favoured by high moisture and moderate temperature (28-30°C) condition (206). Occasionally the pathogen may be seed borne.

Host resistance :

In Madhya Pradesh, India, out of 21 varieties of blackgram, D6-7/1, NP 6, BR 68 and No. 1766 were found to be resistant against leaf infection by *S.rolfsii* (206). Among the greengram varieties, No. 305 has been reported as resistant while Pusa Baishakhi, PS 16, N 412 and L 24-2 exhibited the moderately resistant reaction (276).

Control :

Chemical : Seed treatment with thiram +PCNB (1:1) @ 2.5 g per kg of seed has been recommended (231). Seed treatment with carboxin + captan has also been reported quite effective and recommended for checking the pre-emergence damping off in greengram due to *S.rolfsii* (82).

Though not practical, formaline or methyl bromide when used as fumigants, reduce the disease. Application of 20% PCNB dust @ 25 kg per hactare in the soil at the time of field preparation is also effective.

Cultural : Deep summer ploughing, selection of uninfested field, use of clean and healthy seed, avoiding shade and thick sowing, proper weeding, avoiding mechanical injury to the plant and destruction of the infected crop residues are helpful in reducing the disease greatly.

Biological : *Trichoderma viride*, when applied to seed, checks pre-emergence death of blackgram caused by *S.rolfsii* (229).

2.12. ANTHRACNOSE :

The name "anthracnose" is derived from the Greek word meaning `ulcer', and is appropriate because of ulcer like lesions produced on pods particularly in common bean. The disease affects directly all parts of the host plant except roots. The typical characters of the disease are sunken spots due to drying up and collapse of the cells in the centre and, with the presence of the pink, sticky spore masses that develop during the moist weather, have quite an ulcer like appearance. In general, species of *Colletotrichum* produce the anthracnose symptoms. The disease has been reported occurring on greengram as well as blackgram from several countries including India, Nigeria, Thailand, Philippines, Upper Volta, Zambia, Palmira, Columbia etc. (CMI distribution map No. 177, 231, 249). In India, the disease was reported first from Assam (191). Atleast four species of *Colletotrichum* have been found associated with greengram and blackgram infecting leaves, stem and /or pods seriously in different parts of the world (20, 120, 191, 302, 343, 386, 439).

Nature and extent of damage :

All the species of *Colletotrichum,* associated with greengram and blackgram, cause serious leaf spotting ultimately resulting in `shot hole' symptoms and finally defoliation which affects the yield greatly. Infection of pods directly damages the seed and reduces its germinability. Under favourable weather conditions, the disease spreads very fast and some species may infect the stem resulting in `die back' symptoms and thus damaging the foliage seriously, Since, the disease appears quite early in the season at Jabalpur, it gets sufficient time to damage the crop seriously (308). Pod infection may result in complete loss in yield.

Symptoms :

The parts of the host plant infected and the symptoms produced by different species of *Colletotrichum* differ considerably. Hence, the symptoms produced on greengram and blackgram are described separately for different species of the pathogen:

i. ***Colletotrichum lindemuthianum*** : It is reported to attack both the crops (191) causing true anthracnose (20,130). The pathogen infects seedlings, stem, petioles, pods etc. Brown, sunken lesions are formed on the cotyledons and the young stems of seedlings. During wet conditions, the lesions increase in size and number and young plant is killed after attack. On pods, depressed, black cankers appear, some times with flesh coloured centres. Seeds may also be infected resulting in pre-emergence death. In severe infections, the affected parts may wither off.

ii. ***Colletotrichum capsici*** : The symptoms produced by this species have been categorized as leaf spots and brown blotches (231). The pathogen infects blackgram (386) as well as greengram (37) on which the symptoms are more or less similar. Small, circular, brown spots appear mainly on leaves. Later, the spots enlarge and develop concentric rings giving the target board effect. The concentric rings are dark brown while the in between area is grey/dirty grey. The spots may be sickle shaped, circular to irregular in shape with grey centres and measure 3-5 mm in blackgram and 4-8 mm in greengram. the spotted portion often becomes papery and fall off the leaves producing `shot holes'. Pods of greengram are not infected but dark brown, circular spots appear on pods of blackgram (386).

iii. ***Colletotrichum truncatum*** : It is reported causing serious leaf spotting in greengram (439) and blackgram (302). On blackgram leaves, 2 x 1 to 4 x 4 mm size spots appear which later turn light to dark brown in colour. Spots are circular to irregular in shape with dark brown margins either scattered over the leaf surface or concentrated near the midrib region. The lesions may coalesce forming 4-6 mm, circular to rectangular, necrotic spots. According to Singh and Shukla (450), the disease manifests mainly on the leaves but under severe conditions it may produce small, reddish brown, 1-4 mm diameter spots on petioles; light grey, water-soaked, irregular, later on turning to rustic brown, 10-14 mm diameter spots on stems and small, reddish blotches on pods (Fig. 2.18) which later become light brown to greyish in colour showing numerous acervuli. The leaf spots later form `shot holes' and in severe cases premature defoliation occurs.

Fig.2.18.
Blotches on black gram pods by *Colletotrichum* sp.

iv. ***Colletotrichum dematium*** : It produces leaf spot and blight symptoms on blackgram with the formation of first pair of true leaves (317). Small, water soaked, greenish spots appear on leaves which later enlarge forming light green, irregular patches. When fully developed, they become straw coloured with narrow reddish-brown margins. Infected tissues become thin and papery which shed giving a `shot hole' effect after a spell of sharp showers of rains but narrow band of reddish-brown tissues remains intact. Acervuli with numerous, black, stiff setae can be seen on infected tissues after humid weather.

The symptoms produced by *C. truncatum* and *C. dematium* do not differ much from each other and they are considered under the name *C. dematium* forma *truncata* (425). However, symptoms produced by *C.dematium f. truncata* on greengram and balckgram differ from each other considerably (42). The pathogen produces water-soaked, greenish lesions which assume the shape of horse shoe (Fig. 2.19 and 2.20). Such lesions turn straw coloured with narrow, reddish-brown to dark brown margins. After a few days, necrotic areas shed leaving behind reddish-brown margin intact. In case of greengram, the margin around the necrotic area becomes irregular in shape as a result of extension of the infection on the lower leaves. Usually, such infected portion gives brittle appearance to the spot which results in fall of the small necrotic areas. The spots occasionally coalesce and the acervuli on the necrotic area are not visible to naked eye.

Fig.2.19.
Colletotrichum dematium f. truncata on black gram.

Fig.2.20.
Colletotrichum dematium f. truncata on green gram.

Causal organism :

i. *Colletotrichum lindemuthianum* (Sacc. & Magn.) Bri. & Cav.
ii. *Colletotrichum capsici* (Syd.) Butler & Bisby
iii. *Colletotrichum dematium* forma *truncata* (Pers. ex Fr.) Gr.

On blackgram and greengram, *Colletotrichum lindemuthianum* (Fig. 2.21c) was reported first from India causing anthracnose (191). The pathogen has wide host range. The *mycelium* is branched, septate, hyaline at first but becoming dark with age. The *acervuli* are saucer shaped, subcuticular and become erumpent. On host, few, filiform, brown, 2-4 septate *setae* are formed which are 3.5-12 μ wide and less than 100 μ long (260). The *conidia* are borne acrogenously on short conidiophores and appear pink in mass. They are 1 celled, hyaline, oblong, cylindrical with rounded ends or with one end slightly pointed. Conidia measure 10-20 x 3-6 μ in size and germinate by 1-4 germ tubes. The *appressoria* are sparse, pale to dark brown, clavate or circular in out line, regular and 8 x 6.7 μ in size. The ascigerous stage is *Glomerella lindemuthianum* (Sacc. & magn.) Shear and has been observed in culture but requires confirmation (332).

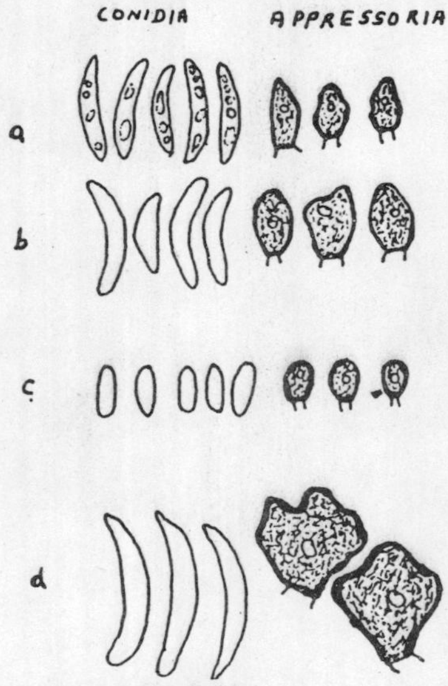

Fig.2.21.
Spores and appressoria of *Colletotrichum* spp. : a. *C. truncatum*, b. *C. capsici*, c. *C. lindemuthianum*, d. *C. graminicola*.

Colletotrichum capsici (Fig. 2.21b) causes mainly leaf spotting and brown blotch symptoms on greengram as well as blackgram (37, 386). It resembles with *C. truncatum* except that the sclerotia are absent and the host range is wide. *Acervuli* are rounded or elongate, 350 μ in diameter. *Setae,* abundant, brown, 1-5 septate, rigid, hardly swollen at the base, slightly tapered towards the paler acute apex, up to 250 μ long and 5-8 μ wide. *Conidia*, falcate, fusiform, apices acute, narrow, truncate base, aseptate, hyaline, uninucleate, 16-30 x 2.5 -4 μ. Appressoria develop in large number, medium brown, clavate to circular, edge usually entire, 6-25 x 4-10 μ in size and forming long, closely branched chains.It is regarded as unspecialized pathogen.

Colletotrichum dematium f. *truncata* has been reported occurring on greengram (439) as well as blackgram (301) under the name *C. truncatum*. (Fig. 2.21a). The cultural characters as wel as the morphology of the blackgram isolate has been described in detail (301, 330). The colonies are dark brown to black with septate, branched, brown and 3-4 μ thick *mycelium* on PDA. The *acervuli* are hemispherical to truncate, conical in shape and measure 150-250 μ. Immature acervuli are greyish white to dull orange in colour but at maturity they turn dark brown to black. The *setae* are dark brown to balck, 6-10 per acervulus, filiform, 1-3 septate, 60-300 μ long and 3-8 μ wide. *Conidiophores* are short, erect,simple, hyaline, non septate and 12.5-20.5 x 2-3.2 μ in size. *Conidia* are hyaline but pinkish in mass, unicellular, thin walled, falcate to lanceolate, bluntly tapering towards both ends and measure 16-20 x 3.0-3.5 μ. The mycelium has greater tendency to form the sclerotial aggregates. According to Saxena and Sinha (301), it seems to be the abnormal isolate of *C.truncatum* as it sporulates slow and has the conidial size at the lower end of the size range which is 17-32 x 3.5-4.0 μ. On blackgram, leaf spot and blight symptoms have been described to be caused by *C. dematium* but the detailed morphology of the pathogen is not given (317). The detailed synonymy of this pathogen has been given by Arx (425) who considered the pathogen under the name *C.dematium* f. *truncata*. Some workers consider the *C. truncatum and C. capsici* as synonyms to *C. dematium* (435).

Colletotrichum dematium is now treated as the big group *C. dematium f. truncata* includes isoloates of the species infecting leguminous hosts. Isolates of *C. dematium* f. *truncata* from blackgram, greengram, horsegram and soybean are reported to differ from each other in regard of their pathogenicity on six leguminous hosts (42) but blackgram and greengram isolates did not differ much except that the adzuki bean variaety HPU 1 was resistant to blackgram isolate but susceptible to greengram isolate. On PDA, both formed dirty white to dark colonies with abundant sporulation but mungbean isolate was darker with leathery colony. The blackgram islolate of *C.truncatum* grows and sporulate best on Kirchoff's medium but acervuli are formed well on all the media tested (330). The incubation temperature 28°C and 6-7 pH of the medium are best for growth and sporulation, though, it can grow at 5-48°C temperature and 2-10 pH. The sporulation takes place at 15-35°C temperature and 3.5-9.0 pH only.

Besides the above mentioned species, *C. graminicola* (Fig.2.21d) has been detected from the seeds of blackgram and is considered to cause the seed rot, and infection of hypocotyl, leaves and pods (165, 343, 435). On blackgram , the *acervuli* are single or in groups; *setae,* numerous, brownish black to black, trichiform, 1-3 septate 57-105 μ long and 3-6 μ thick. Conidial mass, dull white to dull orange; *sclerotia,* mostly present and blackish brown; *mycelium,* whitish orange on seed; *conidia* formed in false heads on tips of conidiophores, hyaline, falcate, one celled, with both ends pointed, 15-18 x 3-5 μ in size (435).

Singh and Srivastava (446) reported that the blackgram and greengram are also infected by *Colletotrichum gloeosporioides*. They found that shape and size of the conidia; presence and position of chlamydospores; shape, size and colour of the appressoria and presence/ absence of the typical sclerotia are the consistant characters for identification of the species of *Colletotrichum*.

Host parasite interaction :

C. truncatum influences the nitrogen metabolism of the blackgram plant. It reduces the arginine, glutamic acid, histidine, lysine, methionine, phenylalanine, thieonine and valine but increases the cystine and proline contents of leaves after infection (331). The protein content of the urdbean seed is reported to reduce by 25% due to the infection by *Colletotrichum* spp. (165, 343).

Disease cycle and epidemiology :

The *colletotrichum* spp. have been found associated with the seed of blackgram as well as greengram (274, 343, 423, 435, 439, 440). The association with seed is external as well as internal and the acervuli are formed in the space between the cotyledons (165). The pathogen can survive on the diseased plant debris also and both are the sources of the inoculum for the primary infection in the field. Further spread of the disease is through the air-borne conidia or the inoculum transmitted through the rains (312). The disease is favoured by the cool and wet weather conditions (326).

The intensity of the mungbean anthracnose has been found negatively correlated with the temperature. The most favourable weather conditions for the disease development are : 26-30°C temperature, 90-100% RH in the morning and 80-93% at noon, rains with a wind velocity of about 13 km/h and the infection of the plants is favoured by overcast 2 and partially cloudy weather 1 but not by clear weather (376).

Higher doses of P_2O_5 (100 kg/ha) are reported to increase the intensity of *Colletotrichum* spp. on blackgram (342) but it increases the yield also. The disease can reach a highly damaging level as it appears quite early in the season in some localities (308).

Host resistance :

Large number of greengram and blackgram entries have been screened for resistance against the anthracnose pathogens of which some possess resistance (Table 2.2).

Most of the commercial varieties have been found susceptible against anthracnose (329). It has been noted that the reaction of blackgram and greengram genotypes against the anthracnose pathogens is not consistant (6). Varieties Pusa 109 and Pusa 115 showed highly resistant and resistant reaction in one season of test but exhibited susceptible reaction next year (377). During the inheritance study, it has been observed that blackgram lines P 7, P27, P 103 and P 115 carry single, dominant, non allelic resistance gene while P 53 carries 2 genes whose interaction is not clear (157). As regards the scoring against the disease, the 0-9 point scale has been suggested (198).

Table 2.2

Resistance in greengram and blackgram entries against anthracnose fungi

Crop	Test pathogen	Total entries	Resistant entires	Refernce
Blackgram	*C.truncatum*	609	Line 112, BR 10 and 6 cultures were MR	(329)
	C. capsici	82	T 65, UPU 79-2-4, UPU 80-5-5, UG 201 PDU 2, 3, 8 and 10.	(265)
	C. truncatum	48	13 lines were resistant including P 7, P 27, P 103, P 115 and P 53.	(157)
	C. truncatum	-	Barabanki local, UH-80-9	(6)
Greengram	anthracnose	-	Pusa 109, Pusa 115	(377)
	anthracnose	-	ML 326	(6)

Grade	*symptoms*
0	No symptoms on leaves
1	Small pin-head size lesions covering 1.0% of leaf area.
3	Small pin-head size lesions covering 1-10 % of leaf area.
5	Lesions big but not coalescing, covering 11-25 % of the leaf area.
7	Lesions on leaves covering 26-50 % of leaf area. Cankers on stem and pod infection.
9.	Lesions on leaves covering 51 % or more of leaf area. Defoliation of leaves; deep cankers on stem and pods. Blighting of plant occurs.

Control :

Chemical seed treatment : Several fungicides have been tested *in vitro* against the commonly occurring pathogen *C. dematium* and *C. lindemuthianum.* The growth of both the fungi was checked by carbendazim, captafol and benomyl (378). Since the species of *Colletotrichum* are seed-borne in nature, it is essential to eliminate the pathogen by fungicidal seed treatment. Thiabendazole and agrosan GN have been found effective in checking the seed-borne pathogen in blackgram (343). Thiram or captan @ 2-3 g/kg seed or carbendazim @ 1-2 g/kg seed were also found effective by Singh and Gurha (326). From blackgram seed, the pathogen can be eliminated by hot water treatment at 58°C for 15 minutes (343).

Chemical spray : Spraying the crop with 0.2 % zineb or ziram at 15 days interval considerably reduces the infection of anthracnose. Application of 0.2 % a.i. benomyl or mancozeb also controls the disease effectively (231). Thiabendazole at 225-315 g a.i./ha. applied at an interval of 14-21 days from the pod set, is also recommended. In Philippines, mungbean anthracnose is reported to be controlled by use of fungicides (86). In Bangladesh, the mungbean anthracnose is reported to be checked effectively by spray of benomyl, chlorothalonil or thiabendazole with an increase in yield up to 79 per cent (35). Against the mungbean anthracnose, spray of carbendazim has been found to be most effective and economic (341) in India.

Cultural : Build up of the disease does not differ much under sole cropping or the intercropping of blackgram with sorghum, soybean or pigeonpea (422) but the disease was observed minimum (35 %) in crop sown on 23rd July as compared to 54 % in case of sowing of 3rd July. It is recommended to follow the proper crop rotation and to destroy the diseased crop debris in the field for reducing the inoculum build up in the area. Procuring the seed from the healthy crop and destruction of weeds also help in minimizing the disease severity on blackgram and greengram.

2.13. POWDERY MILDEW :

'Powdery mildew' or 'white mildew' is a disease caused by fungi belonging to the class Ascomycetes order Erysiphales and family Erysiphaceae, On blackgram and greengram, this disease seems to be reported first from India by Butler in 1918 (55). The disease was reported to be caused by *Erysiphe polygoni* but later other fungal species were also found associated with the disease in different parts of the world. The disease is very widespread and occurs wherever these crops are grown particularly the Indian subcontinant and South-east Asian countries (55, 413). So far the disease has been reported from Ceylon (232), China (371), Japan (391), Nicaragua (189), Philippines (259), Thailand (219), India, Pakistan, Bangladesh, United States of America (249), Iran (107) and several other countries. In India, the disease occurs in almost all the states causing serious damage to these crops specially in Maharashtra (234), Andhra Pradesh (394), Madras and northern state of the country (55), Madhya Pradesh (308), Rajasthan (149), Himachal Pradesh (352) and Uttar Pradesh (199).

Nature and extent of damage :

The disease may cause complete failure of crop due to death of plants if they are infected at early stage. If the infection is late in the season, as is observed at most places, then the disease may adversely affect the shoot length, root length, nodulation and plant growth ultimately resulting in heavy yield losses (323). The damage to the crop plant is caused due to heavy powdery growth of the fungus on the foliage which checks photosynthesis and laso kills the tissues in later stages. Not only the yield but the disease causes reduction in protein content of seed also (225). The extent of damage depends on the stage at which plants get infection and the weather conditions prevailing during the crop growth period. At most places the disease is observed appearing late or the flowering stage. It can reduce as much as 21% yield of greengram when all the leaves are covered with mildew at flowering time (259). A negative correlation has been established between the intensity of powdery mildew and per plant yield of greengram (218). In Taiwan, the powdery mildew is reported to cause 9-50 % yield reduction of greengram and blackgram in farmers fields (31). The disease occurs severely in Taiwan, Thailand and Philippines and the reduction in yield may range from 20-40 per cent (186). The crop is reported to be infected as early as seedling stage of plants particularly in Taiwan, Thailand philippines (186, 249) and parts of Madhya Pradesh in India (218). Under such conditions, the crop may fail completely or there may be very high losses in the yield. The plants may be killed even before flowering and fruiting. The stage of plant at which they are infected differs from place to place. For example In Madhya Pradesh, India, the disease appears last in the crop season at most of the places but in Malwa region it can appear even at seedling stage (218). Similarly in Columbia the disease appears after flowering affecting the yield minimum while it occurs early in Taiwan (249). In India, the losses in blackgram and greengram crops due to powdery mildew are more if they are grown in winter as compared to rainy season (385).

Symptoms :

Symptoms of powdery mildew on blackgram and greengram are similar and have been described in detail by Butler (55). The pathogen attacks all the above ground parts of the plant.

The initial symptoms are faintly dark areas on the lower leaves. These areas develop small, white specks on the upper surface of the leaves which enlarge and coalesce forming bigger patches (Fig.2.22). Later, the disease spreads on the upper leaves also and appears on both the surfaces. Coalescing of spots and enlargement in their size ultimately result in almost complete coating of white powdery fungal mass on leaves, stems, pods etc. Severely affected plant parts get shrivelled and distorted (326, 327). The affected leaves turn yellow, then brown, curl and then dry before defoliation occurs. Pods are not formed; if formed they are malformed, small in size, reduced in number and possess only few seeds which are small, shrivelled and of low quality. The white powdery mass on the host surface later, becomes dirty white in colour. Severely affected pods and stems often become purplish in colour. The disease may create forced maturity (326). The disease is severe at flowering but in cases of early infections, the growth of seedlings is arrested, flowering and pod formation is inhibited and the plants die prematurely.

Fig.2.22.
Powdery mildew (*Erysiphe polygoni*) on green gram.

Causal organism :

1. *Erysiphe polygoni* DC
   Synonym : *Oidium balsami*
2. *Sphaerotheca macularis* (Wall. ex Fr.) Lind.
   Synonym : *Sphaerotheca humuli*

Erysiphe polygoni is the main cause of powdery mildew of blackgram and greengram (55, 96, 129, 234, 256, 270, 326, 394). There has been differences of opinions regarding the identity and taxonomy of *E. polygoni* as it is reported attacking large number of hosts. Isolates from different hosts have been studied in detail in India (239) and have been grouped in to 8 species including *E. pisi* DC ex St Amans which infects members of Papilionatae and differs from *E. polygoni* basically described. Infact, identification of powdery mildew fungi is based mainly on the morphology of their ascigerous stage which is not reported on greengram and blackgram. Only *Oidium* state (anamorphs) is produced on these hosts and it resembles to that of *E. pisi*. The isolate form pea is reported to infect greengram and blackgram (41). Further studies on this aspect are needed and at present the name *E. polygoni* should be considered as valid for the strains occurring on greengram and blackgram.

According to Butler (55) the *mycelium* of the fungus is hyaline, fine or thick, septate, persistant and ectophytic. *Haustorium* develops from the appressorium as narrow tube which penetrates the cell wall and swells up in to a rounded sac in the epidermal cell. *Conidiophores* are simple and arise vertically from the superficial hyphae on the host surface. Each conidiophore bears several conidia in chain produced in basipetal succession. Mature conidia fall off quickly giving powdery appearance of the fungus. *Conidia* are hyaline, single celled, elliptical or barrel shaped and measure 24-36 x 15-17 μ in size.

The sexual stage of the fungus has not been observed on greengram or blackgram but it has been described (55). The *perithecia (cleistothecia)* appear late in the season as small, black specks on mycelium. They are round, 90 μ in diameter, composed of distinct polygonal cells and bear hypha like coloured or hyaline, septate, usually unbranched appendages. Each cleistothecium contains 2-8 *asci* which are ovate, sessile, 46-72 x 30-45 μ in size. Each *ascus* contains 3-8 *ascospores* which are elliptical, non septate, hyaline and 19-25 x 9-14 μ in size. (Fig 2.23)

The fungus is an obligate parasite but has been grown successfully in the laboratory on callus tissues of blackgram and greengram (67). The fungus so grown has been confirmed as pathogenic on both the crops.

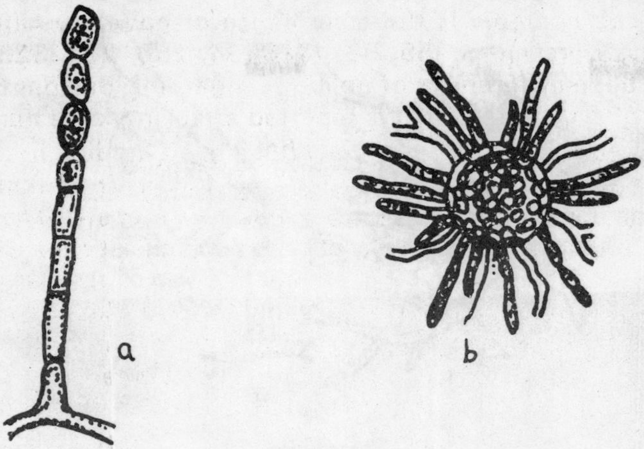

Fig.2.23.
Erysiphe polygoni : a. Conidiophore and conidia b. cleistothecium.

Variability in the isolates of the pathogen has not been studied systematically. However, the host varieties found resistant in one region have been reported susceptible in other region (93). It indicates the possible existence of physiologic races which needs systematic investigation.

Sphaerotheca macularis was reported first from China as *S.humuli* var. *fuliginea* causing powdery mildew of blackgram (371). The powdery mildew of greengram has been reported to be imperfect state of *S. fuliginea* (21, 40, 107). The cucurbit isolate of *S. fuliginea* is reported to be pathogenic on greengram (391). In the state of Himachal Pradesh, India, the cause of powdery mildew of blackgram has been found to be *S. macularis* var. *fuliginae* and the fungus has been described as follows (352).

Mycelium hyaline, septate; *perithecia (cleistothecia)* gregarious, brown to dark brown in colour, globose to subglobose, 65.2-90.8 x 55.9-83.8 μ in diameter. *Cells* irregular, 18.0-41.4 x14.4-27.0 μ; *appendages* variable in number and length, hypha like, flexuous, sparse, brown, mostly simple, septate, some times 1-2 times branched or irregularly bent; *ascus* hyaline, single , subglobose to broadly ovate, sessile, 62.9-88.5 x 58.2-69.9 μ in size. It contains 7-8 *ascospores* which are hyaline, one celled, broadly oblong to elliptical, 18.6-23.3 x 16.3-20.9 μ in size (Fig. 2.24).

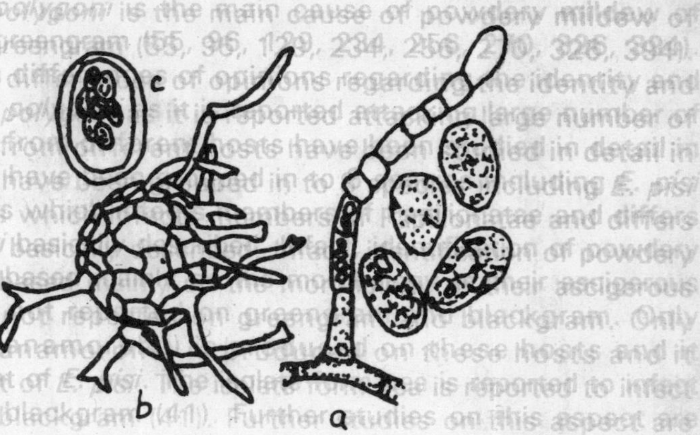

Fig.2.24.
Sphaerotheca fuliginea: a. Conidiophore and conidia, b. Cleistothecium, c. Ascus.

Host parasite interaction :

Infection of host plants by *E.polygoni* changes the metabolism of host. It increases the total nitrogen and free amino acids content in leaves of blackgram but decreases the protein content of seed (225). It influences the sugars and total phenol contents in blackgram as well as greengram (50). In case of greengram leaves infected by powdery mildew fungus, the sugars as well as starch contents are reduced greatly and the reduction in starch is very high if the plant is infected simultaneously with *Oidium* sp. and *Cercospora* spp. (397). The amylase activity is increased in infected tissues.

Disease cycle and epidemiology :

The source of primary inoculum seems to be mainly the conidia formed on annual and perennial collateral hosts on which they may easily survive during the off season. It has also been mentioned (55) that *E.polygoni* can survive on infected leaf tissues in the form of sexual stage in the soil. Though, the ascigerous stage is not formed normally on all the hosts, the disease starts from the lower leaves of the plant due to which the infection foci are suspected in the soil. Once, the primary infection takes place, the secondary spread of the disease easily occurs through wind borne conidia produced on the host surface. Dry and open atmosphere favours release and transmission of the conidia.

The conidia contain sufficient water and can germinate in relatively dry air without free water on the leaf surface (326). In India, the disease on greengram is reported to be favoured by weekly maximum temperature range of 31.34 to 31.67°C with RH 90% during pod formation stage of the plant (207). In the rice based cropping system, the disease appears in epidemic form on winter crop and the RH, wind velocity and rain fall are negatively correlated with the disease spread (1). The maximum disease intensity reaches with in 3-4 weeks after its appearance. In the Philippines, the disease becomes severe in January when monthly means of temperature, RH, wind velocity, total solar radiation and rain fall become 25.6°C, 85%, 1.8 mph, 9976 g cal-cm^2 day and 1.61" respectively (356). Hence, a powdery mildew forecasting system based on the relationship between disease appearance and age of the host has been devised. Accordingly, it is recommended in the Philippines that the mungbean should be planted at the time when vegetative growth does not coincide with the environmental conditions favourable for the disease development.

Alteration of sowing date of the crop may increase or decrease the disease severity to a greater extent. In the Maharashtra, India, minimum powdery mildew incidence is reported on greengram crop sown in third week of August as compared to maximum disease in crop sown in fourth week of July. However, the increase in yield with late sowing is fruitful only in drought prone areas (403). The intensity of powdery mildew on mungbean crop sown in the middle of August is reported to be 58.8% as compared to 95.7% in late July sown crop (207). Under South Indian conditions, the maximum disease occurs on crops sown between 1st to 20th August while it is less in crop sown on 10th September (347). Also, the severity is more on closely planted crop (30 x 5-10 cm) as compared to widely planted one (30 x 15 cm).

The condition of plant is also important in ascertaining the disease severity. The intensity of the disease is much higher on 60-80 days old plants as compared to 40 days old (347). similarly host nutrition may influence the development of powdery mildew. In case of blackgram, optimum nitrogen and high phosphorus (30 kg N +

.0 kg P/ha) minimizes the intensity of powdery mildew when compared with the other regims of nutrients (224). On the contrary, increase in the disease due to higher P_2O_5 application has also been reported but it ultimately increases yield (342). It has also been reported that blackgram plants already infected with blackgram leaf crinkle virus develop more powdery mildew as compared to uninfected ones (226).

The rate of powdery mildew development at early, mid and late stages was studied (218). The `r' was much higher during early stage but decreased progressively with the advancing periods of the disease. These values for early, mid and late periods on blackgram were 0.281, -0.0026 and -0.0005 respectively while on greengram they were 0.278, -0.0042 and -0.0019 respectively. The negative values indicate death and defoliation of severely infected leaves at later stage. Spray of crop with bavistin reduces `r' to a greater extent.

Host resistance :

Resistance to powdery mildew is not commonly observed in greengram (349, 403). However, several entries have been reported resistant in greengram as well as blackgram (Table 2.3). These entries may not be good agronomically and require improvement. For example LBG 17 showed consistantly high resistance to powdery mildew but it did not do well in Madhya Pradesh due to its high photosensitiveness and hence, it is recommended for use in breeding programme (7).

For screening, the blackgram and greengram entries against powdery mildew, two grading scales have been used. Singh et al. used 5 categories of disease grades (323) while Mayee and Datar (198) recommended a 9 point scale for scoring against powdery mildew at growth stages of seedling, flowering, pod formation, and pod maturity.

Since, powdery mildew pathogen is an obligate parasite, maintenance of inoculum for creating epiphytotic on test entries becomes difficult. The screening is done in the field as well as in the laboratory or glass house conditions. Recently, a technique of crop season, are reported to control the disease with an increase in the yield of greengram (257). Not only the spray, even the soil application of benomyl is reported to protect greengram plants from *E.polygoni* up to 70 days (256,257). Another systemic fungicide found effective against powdery mildew as well as certain other foliar diseases of greengram is bavistin, the carbendazim (105, 166, 172, 180, 218, 222, 339, 341, 383, 384, 420). In most cases, 2 sprays of 0.05% bavistin at 10-15 days interval starting from 30 to 40 days after sowing are reported to control the disease and increase the yield at least 33% with a cost benefit ratio of 12.2 (172, 341). Seed treatment with bavistin @ 0.5 g/kg followed by 1-2 sprays of 0.02% bavistin at 30 and 45 days after sowing, control the disease in mungbean very effectively (173). Calixin is another fungicide which is systemic and found effective in controlling the powdery mildew of greengram (339) and blackgram (180).

It is suggested to apply the fungicide well in advance before build up of inoculum in the area (356). It would be better to start fungicidal spray from the day, the first symptoms of the disease are noticed on lower leaves of plant. The rate of disease development may help in suggesting the fungicidal spray to be done (166).

Cultural methods : The cultural practices including manipulation in dates of sowing, plant spacing, fertilizer application etc. may be followed for minimizing the disease. These aspects have been discussed in detail under disease cycle and epidemiology.

Biological control : A fungus, *Ampelomyces guisqualis* Ces. has been reported as hyperparasite on the powdery mildew fungus, *Oidium* sp. on blackgram in India (16). The possibility of its use as biological agent for control of powdery mildew should be explored out.

2.14. RUST :

First report of greengram and blackgram rust seems to be from India (55) and simultaneously from Philippines (282, 283). Later, the disease was reported from Tashkant region of U.S.S.R. (416). It occurs in most of the legume growing areas (CMI description map No. 290) but serious losses to greengram and blackgram crops have not been reported so far from any area.

Symptoms:

Symptoms are more or less similar on greengram and blackgram and have been described in detail (91, 130, 282, 283). In the beginning, tiny, light coloured flecks or spots appear on both the surfaces of leaves. Within few days, spots become dark reddish brown, circular pustules about the size of a pin head. The rusty or cinnamon-brown colour is due to uredo sori containing powdery masses of uredospores. The pustule may be surrounded by a yellow halo. Severely infected leaves turn yellow, then brown as they dry up, shrivel and finally defoliation occurs. Foliage of susceptible varieties is damaged and defoliated to a greater extent particularly in periods of sporadic and intermitant rain fall. It results in serious yield losses. Ilag (130) did not find any other spore stage aside from the uredial stage.

Casual organism :

1. *Uromyces appendiculatus* (Pers.) Unger.
   Synonyms; *U. vignae* Barclay
   U. radiata
2. *Phakopsora pachyrhizi Syd.*

In India, Butler (55) described *U.appendiculatus* infecting several leguminous hosts including greengram, blackgram, cowpea etc. He mentioned *mycelium*, septate, branched, localized, intercellular with haustoria sent in host cells. *Aecia*, not frequent, *aeciospores*, 18-36 x 16-24 μ in size and oblong in shape. *Uredosori* are sub-epidermal and rupture releasing round to ovate, light brown, spiny *uredospores* with 2 germpores. Spores measure 18-28 x 18-22 μ in size. The *telial sori* appear late and are black, warty containing *teliospores* which are subglobose, ovate, elliptical and rounded at the apex with a large hemispherical, hyaline papilla. They are smooth, chestnut brown and measure 24-30 x 18-25 μ in size (Fig. 2.25).

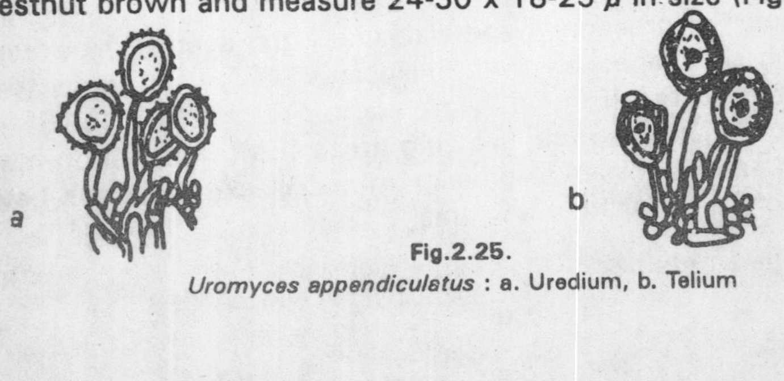

Fig.2.25.
Uromyces appendiculatus : a. Uredium, b. Telium

Ilag described the fungus under the name *U. vignae* from Philippines and observed only the uredial stage (129). He reported *uredia* as subepidermal, erumpent, amphigenous, cinnamon brown on petioles. *Uredospores* one celled with 2, usually supra-equatorial, pores; obovoid or broadly ellipsoid with light golden, echinulate and 1.2 - 1.73 μ thick walls. Uredospores measured 25.4-32 x 18.8-25.0 μ in size.

The pathogen has very wide host range infecting the members of Leguminosae particularly cowpea, horsegram, hyacinth bean, yard long bean, french bean etc (231). The pathogen affects the starch and sugar content of host plants. In greengram plant, it induces the amylase activity reducing the sugar content while starch content and phosphorylase activity remain unaffected (397).

Phakopsora pachyrhizi has been found infecting greengram plants (413). The greengram has been reported to be highly susceptible to soybean isolate of *P.pachyrhizi* in Brazil (284).

Disease cycle and epidemiology :

The teliospores remain viable in crop debris left in the soil which germinate, bear sporidia and consequently infect the new crop in the field. The pathogen is autoecious and all of its four stages occur on the same host plant. Uredospores are the repeating spores and result in build up of the inoculum during the periods at the begining and end of the rainy season and also during the dry season in irrigated plantings. It is favoured by cloudy and humid weather with temperature range of 22 to 28°C. The spores spread by the wind, insects and through contact with animals including man and his implements.

Control :

Chemical : Spray of calixin (tridemorph) at 0.07% concentration has been found to be the best for control of rust. It is followed by Bacor (triadimanol) 0.1%, Bayletan (triadimefon) 0.02% concentractions (209). Baycor 200 EC (biloxozol) has been found effective against rust on blackgram giving 98.8% control and marked increase in yield even with one application as spray (215). Muthusamy and Ragupathy (216) controlled the greengram rust by two sprays of vitavax at 45 days of sowing. The quantity of fungicide was used @ 500 and

750 g/ha in first and second spray respectively. It controlled the disease as well as increased the yield considerably. Amongst the non-systemic fungicides, mancozeb 2-4 kg/ha or zineb 200g a.i./100 l of water applied at 7-14 days is effective (231).

Cultural : Crop rotation should be practiced to check the build up of inoculum in the area. Removal of diseased crop debris and their destruction may help in minimizing the disease in the field. The yield losses can be minimized by adjusting the dates of sowing in such a way so that the incidence of the disease is avoided during the pre-flowering and flowering stages of the crop.

Suitable resistant varieties of greengram or blackgram have not yet been reported.

2.15. FALSE RUST :

A very severe disease of blackgram was observed in Maharashtra, India which was mentioned to be caused by *Synchytrium phaseoli* Patel, Kulkarni & Dhande (235). Later, Payak (244) reported Synchytrium disease of blackgram from Maharashtra and claimed it to be the first record in India. He examined the type specimen of the disease collected by Patel *et al.* (235) and reported that it was not a species of *Synchytrium*. Simultaneously the disease was reported from Agra, India on blackgram as well as greengram (102). Later, the disease was observed in Poona and Khandala (236). However, it seems to be of limited occurrence and is not recorded else where.

Symptoms :

According to Payak (244), only lower leaves of blackgram, well protected from the direct sunlight, are infected and due to improper leaf expansion remain smaller in size showing galls on both the surfaces. Thickening of leaf tissues in the infected region is not observed. Formation of prosorus and the sporangial stages are not observed. Galls on leaves are amphigenous, simple, rarely coalescing and compound appearing as tiny tubercles. Gupta and Sinha (102) observed galls on leaves and stems of greengram and blackgram. The galls are scattered rather uniformly, mostly single, rarely compound, spherical, cupulate and 0.4-0.6 mm in diameter.

Causal organism :

Patel *et al.* (235) first reported *Synchytrium phaseoli* Patel, Kulkarni & Dhande sp. Nov. on blackgram. A critical examination of this specimen did not prove it to be the species of *Synchytrium* (244) but a species of *Protomycopsis* (242). Hence, Payak (244) while reporting this disease from Maharashtra, claimed it to be the first record in India and proposed the name of the pathogen as *Synchytrium ajrekari* Payak & Thirumalachar Sp.Nov. Simultaneously, the disease was described from Agra, India and the causal organism was named as *Synchytrium phaseoli radiati* Sinha & Gupta Sp. Nov. (102).

S. ajrekari : *Hypnospores* spherical, filling the host cells, golden brown, 114-266 μ in diameter, surrounded by a hemispherical mound of thin walled tissue above; thick walled, wall 8-11.5 μ thick. Found on lower leaves of blackgram (244).

S.phaseoli-radiati : Resting *sporangia* solitary in epidermal cells of host, spherical, smooth, dark brown, measuring 165-200 μ (mean 184 μ) in diameter with an epispore 13-16.5 μ thick. Reported on leaves and stems of blackgram and greengram (102).

S.phaseoli was described on blackgram (235) but later examination of this material proved it to be a species of *Protomycopsis* (242). It possessed smaller spores (18-26.6 μ) with warty wall and were present intercellularly in host tissues. Such characters are not found in *Synchytrium*.

Disease cycle and epidemiology :

Since, the disease is not wide spread in nature, it did not attract the attention of research workers. *Synchytrium* spp. have been reported on cowpea and winged bean. They are reported to perpetuate in diseased crop residues, spread through wind and their development is favoured by wet weather (231). It may be true with the *Synchytrium* spp. attacking greengram and blackgram also.

Control :

No specific work is reported on this aspect. As in case of cowpea, the disease is expected to be reduced by destroying infected plants at early stage and spraying the crop with mancozeb 0.25% with the start of disease. Destruction of infected crop residues after harvest will help in checking the perpetuation of the pathogen.

2.16. ANGULAR BLACK SPOT :

A severe disease of blackgram was observed in Maharashtra, India, the cause of which was wrongly identified as *Synchytrium phaseoli* (235). After a careful examination of this specimen by Pavgi and Thirumalachar (242), the causal organism was identified as *Protomycopsis patelii.* The disease was also observed in Banaras, India. On the basis of survey conducted during September and October over several years, it was revealed occurring on greengram as well as balckgram in the states of Uttar Pradesh, Madhya Pradesh, Rajasthan, and Maharashtra (110). In Andhra Pradesh, India, the cause of the disease on greengram has been identified as *P.phaseoli* (251) which can cause yield losses up to 40 per cent (59). The disease causes more losses in blackgram as compared to greengram (6). The disease is not very wide spread and occurs on greengram, blackgram, cowpea and common bean. It is restricted to India, Tropical Africa, Brazil and Jamaica only (231).

Symptoms :

The disease is characterized by the appearance of purple-black, opaque and typically angular spots mostly on lower leaves (Fig.2.26). The spots are often mistaken fo *Cercospora* leaf spots occurring on these crops during the same period. The *Cercospora* spots are greyish brown in the centre. In case of angular black spot disease, the spots never disintegrate and are generally bound by the veins (110)

In the field, the infection sites can be identified during the morning hours as minute diffused chlorotic areas on the leaflets. Initially, 3-5 pale green spots appear mostly on lower leaves frequently comming in contact of soil. Lesions gradually deepen in colour and enlarge in size. Centres of these lesions turn purple, spreading radially. Finally spots turn black to opaque, remain angular as they are limited by veins, 3-4 mm x 6-8 mm in size without any chlorotic area around them. In severely diseased crop, 70 to 100 spots per leaflet may be observed. Lesions are pale brown on lower surface and no malformation of leaves is observed.

Fig.2.26.
Angular black spot on black gram.

Under saturated atmosphere, spots show tendency to enlarge 'ith a yellow chlorotic area around them. Under dry conditions the sions remain minute and pale green while during humid weather, ey are well developed, black and angular resulting in premature ellowing and defoliation of leaves. Number of spots are more in ackgram as compared to greengram.

ausal organism :

Protomycopsis phaseoli (Patel *et al.*) Ramakrishna & Subra-anian.

In beginning, pathogen of this disease was wrongly identified as *Synchytrium phaseoli* patel *et al.*(235). Later, Pavgi and Thirumalachar transferred it to the genus *Protomycopsis* on the basis of intercellular mycelium in the mesophyll of host tissues and rough walled, terminally formed dark brown chlamydospores (242) and named it *P. patelii* Pavgi and Thirumalachar. Prasad *et al.* (252) however, discussing the taxonomy of the genus *Protomycopsis*, considered only two species depending on similarity in symptoms and morphological characters of chlamydospores. They are *P. phaseoli* attacking *Phaseolus radiatus, P. mungo, Vigna sinensis, Crotalarea triquelea* and *Smitha sensitiva*, and *Protomycopsis ajmeriensis* parasitizing *Sesbania aculeata* and *S. aegyptiaca*. Hence *P. patelii* should be considered as a synonym to *P. phaseoli.* Infact the taxonomy of *Protomycopsis* itself has been very doubtful due to lack of precise knowledge about its developmental cycle. However studies on cytology of development and germination of chlamydospores suggested it under class Ascomycetes (119).

The morphology of the fungus *P.patelii* has been studied in detail (110). The *mycelium* is intercellular, septate and branched developing numerous dark to reddish brown resting spores in mesophyll. *Resting spores (chlamydospores)* are formed terminally on the hyphal branches. Young spores are thin walled, hyaline and pyriform, turning deep brown and warty as they mature. Matured spores are globose to oval, thick walled, covered with pyramidal warts and measure 22.5 x 32.5 *µ* in diameter with a 2.5-3.4 *µ* thick epispore. Germination of chlamydospores on slide mounts shows a close similarity with *Protomyces* in the formation of an endosporic vesicle, meiotic nuclear division in the peripherally arranged diploid nuclei and tetrad formation of the haploid (endo) spores which exhibit a clear alteration of generation (115, 119). The chlamydospores and edosporic vesicle, after germination, can well be compared with a synascus as a diploid body in which meiotic division of diploid nuclei takes place followed by mitotic division.

The germination of chlamydospores is facilitated by a brief steep in acidified water which renders the thick episporium permeable and hastens the germination (115). The external nutrition and light factors are unimportant but the germination is favoured by a temperature of 25-30°C and the pH 3-7 (114).

The fungus can be grown on nutrient culture media and *in vitr* levelopment of germinable chlamydospores is possible using special utrients (240). *In vitro* development of chlamydospores is nearly imilar to that *in vivo* except the parasitic relations in the later.

Out of 23 plant species inoculated with *P. patelii* only lackgram, greengram and cowpea were found susceptible (118).

listopathology :

The chlorotic zones formed aroud spots during the saturated tmosphere, contain the actively growing intercellular mycelium amifying in the mesophyll. The opaque black portion of the spot ontains chlamydospores in the inter spaces of mesophyll cells (110). 'he chlamydospores are formed from the terminal sporogenous cells f the permeating hyphae with in the infection epicentre (116). No ascular infections are noted and haustoria are not formed (116). 'he pathogen does not incite any malformation in host tissues articularly in the members of Leguminosae and Compositae.

isease cycle and epidemio. gy :

The pathogen perpetuates through desiccated but viable, heat esistant chlamydospores in host tissues. The heat resistant ıycelium formed more chlamydospores *in vitro* in soil and probably oes the same under natural conditions. Moist heat reduces the urvival(113). The pathogen can survive in crop debris for atleast 2 ears but it is not viable if burried in the soil (231). Due to survival in oil the lower most leaves of plant which are in contact or nearer to oil are infected first. Further spread of the disease is through wind. 'he diploid conidia produced on the mycelium in artificial culture are eported to be expelled forcibly and carried away by the wind current) incite foliar infection at sites of their lodging on leaflets (241). The ermination of chlamydospores is favoured by temperature of 25-0°C (114). The saturated atmosphere enhances the disease while ry conditions restrict the size of spots on leaves as well as spread f the disease.

lost resistance :

Haware and Pavgi (117) screened 13 varieties each of reengram and blackgram against *P.patelii* under artificially ıoculated conditions. In case of blackgram, only 2 varieties were highly resistant while in greengram three were immune and six a highly resistant. Recently greengram varieties LGG 450, LGG 421 LGG 407 and Pusa 105 have been identified as the resistant source againt *P. phaseoli* and recommended for use in breeding programm (59, 220). The inheritance of resistance has been reported to b governed by a single dominant gene in LGG 450 x PDM 54 (220) The author screened 500 collections of blackgram at Sehore, Indi and observed great variability in regard to the reaction against *P. phaseoli* (unpublished data). The line JU 571 was free from th disease while JU Nos. 436, 452, 457, 468, 485, 500, 532, 536 538,543, 545, 555. 562, 569, 572, 574, 592; PLU Nos. 58, 466 495, 623, 1288 and varieties UG 5, P 21, KC 70 and CO 2 wer resistent. Several entries including some commercial varieties lik Pant U 19, T 9, JU 78-4, JU 77-41, RU 2, NP 24 and Pant U 3 were tolerant to the disease.

Control :

Chemical : During the *in vitro* tests, all the copper fungicides an dithiocarbamic acid derivatives at 0.2% concentration caused lysis c the vesicle of germinative chlamydospore of *P. patelii* (111). ziram PCNB and ceresan were ineffective. Aureofungin inhibited spor germination at 120 µg/ml. In a field trial (112) spray with blitox 5 was most effective at lower dose while aureofungin and othe fungicides were effective at higher doses. Seed treatment wit carboxin 5 g/kg seed and spray of crop with benomyl 0.55g/l i recommended (231).

Cultural : Crop rotation and destruction of infected crop debris hel in minimizing the disease.

2.17. CERCOSPORA LEAF SPOT :

Leaf spot of greengram, cowpea and mothbean caused b *Cercospora cruenta* was first reported from India in 1918 by Butle (55). The disease on blackgram was reported later (353, 434 Greengram and blackgram crops are attacked by *Cercospor canescens* (372) and certain other species of *Cercospora* but firs two are found commonly and cause severe leaf spotting resulting i heavy yield losses. The disease is very wide spread and has bee reported from America (246), Bangladesh (10,30), China (372 Indonesia (354), Malaysia (2), Philippines (405), Somalia (92

Sri Lanka (402), Thailand (219), Venezuella (212) and several other countries in South East Asia (205, 249). In India, the disease occurs in most of the greengram and blackgram growing states particularly, Punjab, Haryana, Rajasthan, Uttar Pradesh, Bihar, Madhya Pradesh, Orissa, Himachal Pradesh, Maharashtra, Assam etc.

Nature and extent of damage :

Cercospora spp. produce large number of spots on leaves which may coalesce causing death of leaf tissues and premature defoliation. It, thus reduces the leaf area index (388) which adversely affects the yield. Seeds produced from the infected plants are shrivelled, smaller, with poor quality and reduced germination (441). The disease is more destructive in humid tropics and causes much damage to the crop grown in rainy season (380). The disease has been described as a threat to greengram production in Indonesia, Philippines, Thailand and Columbia (250). It appeared in epidemic form in Uttar Pradesh state of India in 1934 and 1941 (452) and causes serious losses in Himachal Pradesh (103), Madhya Pradesh (6), Orissa (435), Punjab, Haryana etc.

The reduction in yield due to *Cercospora* leaf spot depends on how early plants are infected (249). In greengram, the yield reduction may be 23% if 75% foliage is killed (258). In Taiwan, the disease is reported to cause 37% losses in greengram yield (231). The disease is more severe if wet condition prevail in Philippines and may cause up to 47% losses in greengram yield during warm wet season (186). In India, the yield losses due to the disease have been reported proportional to the disease severity at 50% flowering which is true for high and low progression of disease but not for intermediate progression (366). Increase in disease severity after 50% flowering does not alter the yield much. The area under the disease progress curve is better correlated with yield loss while infection rates are the poor indicators of yield loss due to variation in the onset of disease epidemic. The pathogen *C. cruenta* is widely distributed but in Punjab *C. canescens* causes much damage to greengram crop (230).

Symptoms :

The characteristic symptoms are formation of leaf spots with brown to greyish centres and reddish border (Fig. 2.27). In severe cases, premature defoliation may occur. The petioles, stems and

Fig.2.27.
Cercospora leaf spot on green gram (early stage).

pods may also be infected. The symptoms produced by different species of *Cercospora* may differ slightly from each other.

C. cruenta infects greengram as well as blackgram producing typical leaf spots around flowering stage of the crop (14, 55, 96, 405). The spots on leaves are usually angular being limited by veins (55) but Wells (405) mentioned spots to be circular in shape and not angular. However, after coalescing they may form big spots (11 mm) which may be limited by veins. The spots are larger in size (5-7mm) than those produced By *C. canescens*. They are tan to grey, surrounded by prominent, reddish purple border. Some times centres break away exhibiting `shot hole' symptoms. Severe attack causes premature defoliation after drying of leaves. Stems and pods are also infected (353, 405). On pods, symptoms range from few spots to solid blackening and killing of entire pod (405). Younger pods are not infected. On stems, the spots are enlarged and elongated which become dark and finally purplish. In severe cases, the seeds within the infected pod become shrivelled and dark in colour.

C. canescens is reported to infect both the crops and the symptoms have been described in detail (14, 96, 231). Numerous, small, circular to semicircular, pale brown, 1-3 mm size spots with reddish brown border are produced on leaves. They may coalesce forming big irregular lesions, some times covering larger leaf area finally causing blighting symptoms and defoliation. With the advancing age and the spore formation, the spots become silvery to grey in centre (Fig. 2.28). The fruiting is amphigenous and the spots are formed mainly on leaves. In severe cases, stems and pods may also be infected (284).

Fig.2.28.
Cercospora leaf spot on green gram (advanced stage).

C. kikuchii was reported on greengram leaves and stems from Simla, India (62). The pathogen has been found associated with seeds of greengram as well as blackgram and causes serious leaf spotting on both the crops (422, 437). On leaves, the lesions are angular to irregular in shape, 2-11 mm in diameter, liver-brown to vandejke brown, later tawny and even olivaceous grey (62). They are elliptic on pods and stems. According to Grewal *et al.* (96), spots are subcircular to irregular, pale brown with tan or grey centres and measure 3-6 mm in diameter. They may coalesce forming big lesions on leaves.

C. dolichi is prominent on rabi greengram and blackgram (201) and produces reddish brown blotches (96).

The symptoms produced by *Cercospora* spp. may be influenced by several factors. The lesion size vary depending on the isolate of the fungus and host variety (129).

Causual organisms :

1. *Mycosphaerella cruenta* Latham
   Conidial state :*Cercospora cruenta* Sacc.
   Synonyms: *C. phaseolorum* Cooke
   C. vignae E &E
   C. lussoniensis Sacc.
   C. phaseoli D & B
   Pseudocercospora cruenta (Sacc.) Deighton

The fungus infects greengram as well as blackgram. On greengram, it was observed as early as 1903 (395) in India. Later, it was described by Butler (55). The pathogen has been described from Philippines also (405). The perfect stage of the fungus was described and named as *Mycosphaerella cruenta* (434) but it does not occur commonly.

Conidiophores emerge in clusters through stomata. They are simple, furcate, subdenticulate, dilute olivaceous upwards, septate or non septate and 50-100 x 4-6 μ in size. The *conidia* are subhyaline or olivaceous, elongate, tapering above, straight or slightly curved, acicular - obclavate with 2-8 septa and 50 - 140 μ x 4-5 μ in size (Fig.2.29). However, the conidia produced under high moisture conditions may be as long as 95-200 μ with a maximum of 14 septa

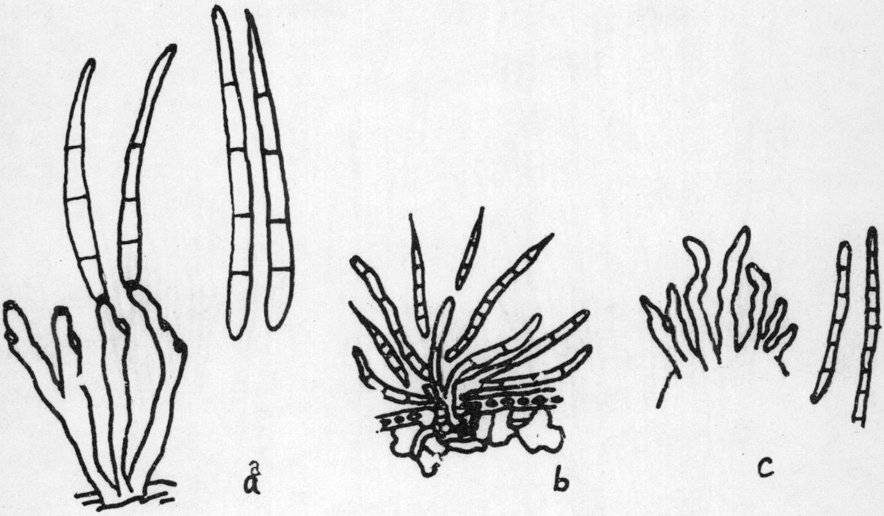

Fig.2.29.
Cercospora cruenta : a, b, c. Conidiophores and conidia in different fascicles of conidiophores.

(129, 405). *C. cruenta* differs from *C. canescens, C. caracallae* and *C.columnaris* in having coloured conidia, dense fascicles and lighter coloured conidiophores (395). The cowpea isolate of *C. cruenta* does not infect greengram and blackgram (396).

2. *Cercospora canescens* Ellis & Martin
Synonym : *C. vignicaulis* Tehon

It is reported to infect greengram as well as blackgram (96, 129, 353, 372). The *stromata* are slight, some of the fascicles dense; *Conidiophores* mostly straight, geniculate, 1-7 septate, rarely branched, pale to medium dark brown, 20-175 x 3-6.5 μ in size, rarely up to 250 μ long. The *conidia* are acicular, straight to variously curved , hyaline 3-20 septate with a truncate base and narrow tip. They measure 50-225 x 3.5-5 μ (average 142 x 4.5 μ) in size (Fig.2.30). The size of conidiophores and conidia differs greatly (63, 96, 129, 353, 395). The fungus infects several other leguminous hosts.

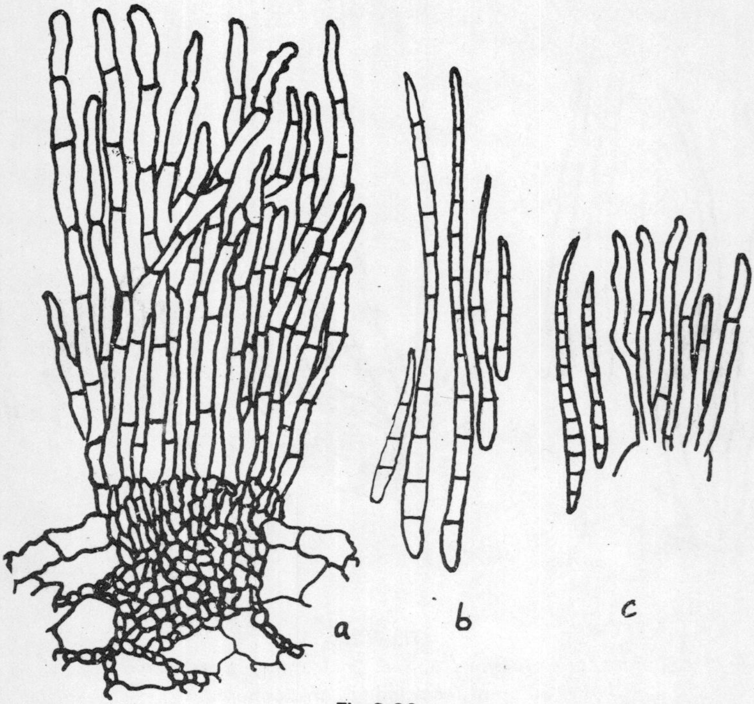

Fig.2.30.
Cercospora canescens : a. Dense fascicle of conidiophores, b. Conidia, c. Conidia and conidiophores with less dense fascicle.

3. *Cercospora kikuchii* Matsumato & Tomoyasun.

The fungus attacks soybean but it is reported to infect greengram in India (62). The fungus has been isolated from seeds of these crops and causes serious leaf spotting (422, 437). However, it does not occur commonly in nature. The *conidiophores* occur in dense fascicles, medium dark brown, multiseptate, 45-220 x 4-6 μ in size. In India, *Mycosphaerella phaseoli* has been reported as the perfect stage of *C. kikuchii* (62). It has been observed on mungbean having *perithecia* 105-126 μ in size with an ostiole 28-35 μ. The *asci* are 40-52 x 6-7 μ and the *ascospores* 8-14 x 3-4 μ in size. Although no organic connection was established between the conidial state and the ascigerous stage of the fungus, the presence of both in the same spot and perithecia often bearing conidiophores resembling to *C. kikuchii* suggests that they are the two stages of the same fungus (62).

4. *Cercospora dolichi* Ellis & Everhart.
Synonyms: *C. vignae* Rac.
C. raciborskii (Rac.) Mats. & Nag.
C. vignae - sinensis Tai & Wei.
C. neovignae Yam.

In India, the *Dolichos lablab* isolate of the fungus was reported to infect greengram (201). Chupp (63) also described *C. dolichi* attacking greengram. *Stromata* not well developed, fascicles dense; *conidiophores* olivaceous, 20-35 x 4-5 μ, simple or slightly denticulate, continuous or 1-2 septate; *conidia* hyaline, 3-5 or multiseptate, obclavate, cylindric, 50-100 x 3.5-4 μ in size. It is different from *C. cruenta* which has narrow and slightly branched conidiophores, fascicle loose to dense, conidia with obconic base and subactue tip. These species can be separated in culture (Fig. 2.31).

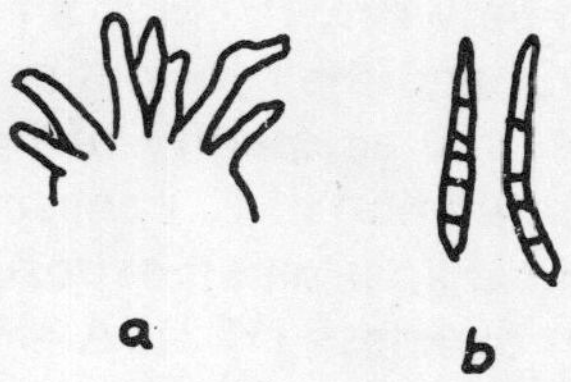

Fig.2.31.
Cercospora dolichi : a. Conidiophores, b. Conidia

5. *Cercospora caracallae* (Speg.) Chupp

It has been found attacking greengram (63). Detailed symptoms have not been described.

6. *Isariopsis griseola*
Conidial state : *Cercospora columnaris* Ellis & Everhart.

The fungus has been observed attacking blackgram in Somalia by Golato (92).

However, under natural conditions the greengram and blackgram have been found attacked mainly by *C. cruenta* and *C. canescens* at most of the places.

Physiology of pathogens : Most of the *Cercospora* spp. are slow growing and their poor sporulation poses problem in building up of the inoculum for creating artificial epidemic. The greengram isolate of *C. canescens* grows and sporulates on PDA but sporulation is best on carrot leaf juice oat-meal-agar medium (205). The fungus grows well in non synthetic media of which Richard's medium is good (161). The sporulation is best at carrot leaf decoction. Glucose or sucrose as carbon source, nitrogen through organic sources particularly alanine or nitrate nitrogen are best for growth as well as sporulation of *C. canescens*. Maximum growth occurs at 22.5 : 1 ratio of C:N (31.5 g carbon and 1.386 g nitrogen per litre) but sporulation is good at low carbon level (5.25 g per litre). Thiamine 10 ppm and Zn 1 ppm support good growth and fair sporulation.

For growth of *C. canescens*, temperature range of 10-35°C (25°C optimum) is favourable while for sporulation, 25-30°C is good. Number of conidia formed between 15-28°C are positively correlated with the temperature (205) and their formation is greatly promoted by light. The quantity as well as quality of light is important for sporulation. Abundant sporulation of *C. canescens* occurs on mung bean leaf decoction oat-meal agar when exposed to c 2500 lux and fluorescent light but not in darkness (183). A pH level of 5.9 is best for growth and sporulation of the pathogen (205). The conditions required for maximum sporulation are not same as for good growth *in vitro.* Removal of the aerial mycelium by brushing with the sterile water, enhances the conidial production on oat-meal agar.

The greengram isolate of *C. dolichi* requires thiamine (1-100 ppm), pyridoxin (100 ppm) and D-biotin (5-10 ppm) partially and Fe and Mn (0.01-0.1 ppm), Zn (0.1-1.0 ppm), Cu (0.01-1.0 ppm) and B (0.0001 ppm) for its growth. Their defficiency affects the fungal growth qualitatively as well as quantitatively (277).

Host parasite interaction :

Infection of greengram by *Cercospora* spp. and *Xanthomonas* sp. causes rupture of epidermal, palisade and spongy parenchymatous cells; increases the intercellular spaces. It results in the degeneration of cell wall, collapse of cells and disorganization and scattering of cytoplasmic contents (445, 447). The mycelium of *Cercospora* is traceable in leaf tissues.

Infection of greengram with *Cercospora cruenta* causes depletion of starch and sugars with concomitant increase in amylase activity. It also increases phosphorylase activity. The rust incidence is reported to reduce the damaging effect of *C. cruenta* on starch content (397). Irrespective of the sequence of inoculation *C. cruenta* develops high disease intensity as compared to *Xanthomonas campestris* pv. *vignae-radiatae* on one month old plant of greengram. The total sugars, total proteins and phenols have been detected more in plants inoculated by bacterium alone as compared to bacterium + *Cercospora* or *Cercospora* alone. Protease activity is recorded more in case of *Cercospora* alone than in combination or bacterium alone. Cellulase activity is reported similar in all cases. *In vitro*, fungus produces more protease and cellulase than bacterium. *C. cruenta* produces cercosporin *in vitro* and *in vivo* (446).

Disease cycle and epidemiology :

Some of the *Cercospora* spp. particularly *C. kikuchii*, *C. cruenta* and possibly *C. canescens* have been found to be associated with the seeds of greengram and blackgram (300, 437, 441). Hence, the source of the primary inoculum seems to be the seed, specially the infected cotyledons attached with the seedlings or lying on the soil surface in the field or the diseased plant debris of previous crop from where the inoculum spreads through splashing rains (312) or the wind. The secondary spread of the disease is mainly air - borne. however, the effective transmission of the pathogen by the air, may be influenced by several factors. The conidia of *C. canescens* loose their viability within 6 h in simulated air - borne conditions (48). Heavy rains are likely to reduce the concentration of the dispersed conidia within the crop canopy by washing down the spores from the foliage.

After transmission, the further development of disease is greatly influenced by environmental and cultural factors. In Philippines, warm-wet conditions are reported to favour the disease (186). In India, *C. cruenta* is reported severe on blackgram at 17.8-30°C temperature with rainfall over a wider period of growing season while *C. canescens* on greengram occurs severely at low temperature (14°C) and 85% RH (284). It seems that the mean temperature levels of 22.5-23.5°C, the RH between 77-85%, sunshine of 5 h per day and more number of rainy days are more conducive to the disease development (365). Intermitant rains and fluctuating RH are favourable. Such conditions are likely to prevail when rainfall occurs mostly during nights and the days are generally bright.

The sporulation and the spore release is favoured by high RH and a temperature of 25-30°C (230, 275). Light is also an important factor for this purpose (183, 257, 275). The sporulation of *C. canescens* on blackgram increases rapidly up to 7 days and then it slows down. A temperature range of 15-40°C (25°C optimum), RH above 81% and complete darkness are favourable for spore production (49). Spore germination is favoured at 25°C and 100% RH or free water (230, 355). White light and a pH of 5-7 favours the conidial germination and darkness inhibits the germ tube elongation of *Cercospora* spp. on greengram (47). It is why the rainy or dewy nights with suny days are conducive for disease development while the continuous dry spells or continuous or heavy rains are not so favourable.

The quantity of spores formed, their dispersion and the speed of their germination determine the comparative severity of the diease caused by that pathogen. The *C. canescens* appears late than *C. cruenta* on blackgram but becomes predominant later on. It is due to the high speed of spore germination at 25-30°C and 3-4 times more spores produced by *C. canescens* in leaf tissues (355). In case of *C. cruenta* spore germination is higher at 35°C. In both the cases the spore production is favoured at 30°C (15-40°C). For creating artificial epiphytotic of the disease on greengram, spray of 2000 spores/ml is required at flowering stage onward (205). The spore germination in both the cases can be induced by phenyl alanine, glutamic acid and arginine (230). The environment greatly influences the reproductive structures of *Cercospora* spp. (85).

The host factors also play an important role in developing the disease. Age of plant is one of them (275). On blackgram, the disease starts appearing when plants are 50-55 days old but becomes severe at the age of 80 days. In Bangladesh also the disease is reported to appear late (423). The greengram plants with more than 2 C_1/RS ratio are more susceptible to *C. canescens.* It has been reported that the nitrogen and phosphorus applied to green gram, increase the disease severity but potash decreases it greatly (346), though, application of P_2O_5 ultimately increases the yield also (342).

Among the cultural practices, the date of sowing is important as it determines, whether the susceptible stage of the crop will coincide with the favourable environmental conditions or not. In India and Philippines, delayed sowing of both the crops are reported to show great reduction in the disease severity (103, 284, 365). On the other hand in Malaysia, the delayed sowing (after May) causes severe incidence of *C. canescens* ultimately reducing the leaf area index greatly (388)

Host resistance :

Sources of resistance to Cercospora leaf spots have been identified in greengram as well as blackgram (Table 2.4). It is necessary to identify additional good sources of resistance and use them for breeding high yielding disease resistant commercial varieties. The functional resistance to Cercospora leaf spot has been mentioned by Ahmad (10) who attributed less infection of variety aurea to smaller pore size and number of stomata per unit leaf area. A positive correlation between the resistance in greengram and stomatal size, cuticle thickness, nitrogen and protein gradient and high total phenols has also been observed (137). The difference in isoperoxidase pattern has been observed among the varieties of blackgram and greengram resistant to *Cercospora* spp. P x 4 and P x 5 is associated with resistance. Peroxidase polymorphism may, therefore be used as a selection index for resistance. The situation process should be undertaken during vegetative and pod bearing stages when P x 4 and P x 5 are detectable (202, 203). The resistance is reported to be governed by a single dominant gene (379, 380) and there is no linkage between resistance and pigmentation of stem and petals of greengram variety PS 7.

For screening the entries against *Cercospora* spp., artificial epiphytotic is required which can be created by spraying the conidial suspension (2000 conidia/ml) on test material at flowering stage onward (205). The scoring against the disease has been recommended in various ways by different workers of which following are important:

1. Scale used by Rath and Grewal (275) :

a. Disease intensity-

	Intensity	Reaction	Numerical score
i.	Few, under-developed lesions	R	0
ii.	Many, under-developed lesions	MR	25
iii.	Few, developed lesions	MS	50
iv.	Many, developed lesions, leaves green	S	75
v.	Many, developed lesions, leaves yellow	HS	100

b. Disease Severity = 0.01 x Plants affected (%) x disease intensity.

2. Scale suggested by Mayee and Datar (198) :

Grade	Description of symptoms
0	No symptoms on leaves.
1	Small pinhead size, brown lesions on leaf covering less than 1% leaf area.
3	Small, round, brown lesions on leaf covering 1-10 % of leaf area.
5	Lesions on leaves are bigger, covering 11-25 % of leaf area.
7	Lesions enlarging and coalescing, covering 26-50% of leaf area, shot hole symptoms produced.
9	Lesions coalescing, covering 51 % or more of the leaf area, shot hole symptoms produced, lesions also on pods.

At growth stages : Seedling, flowering, pod formation and pod maturity.

Table 2.4
Greengram and blackgram varieties/lines identified as resistant against *Cercospora cruenta/ Cercospora canescens*.

Crop	Entries	Entries identified as resistant	Reference
Greengram	730	OB 41, P Nos. 48-68, 273, 546-68, 362-68, 364-68, 11150, 4470-2, 12-14/3-1-1-1-1	(319)
	-	LM 162	(93)
	648	Acc. Nos. 354, 375, 378, 661	(275)
	130	MB 53 highly resistant	(127)
	-	ML 9, ML 56, ML 70-10, 11157	(137)
	29	None was resistant, ML 4, ML 9, Varsha, Local and RRS Pusa, less susceptible	(160)
	-	EC 27087-2, 2761-3, ML 1	(380)
	435	15229	(90)
	-	CES ID-21 (Pag-asa)	(257)
	-	LM 113, 168, 170, 171, 404	(440)
	27	6 entries resistant including 2273, 2773	(186)
	4380	ML Nos. 234, 268, 269, 272, 274, 338, 360, 371, LM Nos. 113, 687, 695, 696, 744, MUG Nos. 125, 130, 132 to 136, 138, 139, 145, 170, 173, M 13, Tarai local	(323)
	-	11/99, 11/395, 16-303-20-3-q	(443)
Blackgram	183	1815, PLU Nos. 2A, 14/8, 291, 13/10, 445, 442, 452, 61	(106)
	12	JU 78-16, Pant U-30, No. 55 (MR)	(451)

Control :

Chemical : The best control of the disease has been obtained by 2 sprays of the crop with 0.05 % bavistin (carbendazim) at 30 and 45 days after sowing. It controls the disease and increases the yield to an economic level (95, 150, 190, 321, 344, 348, 362, 383, 384). The fungicide recommended for each spray is 500 g/ha in 1000 l of water but a quantity of 250-300 g/ha has been found equally effective (93, 95) with 50% increase in yield. Seed treatment with bavistin @ 1 g/kg of seed followed by one spray of bavistin 0.05% at the time of disease appearance control the disease effectively increasing yield upto 3.6 q/ha (11). On the other hand, a complicated spray schedule involving seed treatment and 4 sprays of bavistin in combination with other fungicides has been recommended by Thind and Kumar (381, 450) which seems neither practical to Indian farmer nor economic for the crops like greengram and blackgram. The bavistin, after absorption by leaves, persists there for 15 days and the maximum accumulation in greengram leaves is detected after 48 hours of spray (381). Bavistin (0.05%) when sprayed on leaves, inhibits spore germination of *C. canescens* 100 % up to 15 days. The effect declines afterward indicating that the bavistin can be sprayed on greengram at an interval of 20-25 days (12). However, there are some draw backs of the bavistin. Spray of greengram crop with 0.07% bavistin resulted in control of cercospora leaf spot but it increased the maturity period of crop by 10 days (284). Hence, it is recommended to replace the last spray of bavistin by Dithane M 45. Bavistin is reported to control cercospora leaf spot but it enhances the intensity of bacterial leaf spot (451). Another factor is the high cost of bavistin due to which use of copper fungicides including 5:5:50 Bordeaux mixture has been recommended by some workers (191, 199, 320).

Benomyl also reduces the leaf spotting and defoliation increasing the yield of greengram (151, 324, 362). It is effective at 30 g/ 100 l when sprayed twice at an interval of 10 days (257). It inhibits the spore germination and the mycelial growth of *C. canescens* and controls the disease development on inoculated plants (162).

Though, the fungicides can reduce the disease development, they are unlikely to be technically or economically viable at the farmers level particularly in the high rainfall regions.

Cultural : Since, *Cercospora* spp. have been reported on seeds of both the crops (300, 422, 437, 438, 441), procuring seed from the disease free crop and its treatment with thiram 2.5 g or bavistin 1 g/kg of seed will help in reducing primary inoculum of the pathogen.

As early as 1924, Wells (405) recommended use of resistant lines of greengram and sowing at the time when the crop is likely to escape the heavy rains for controllig the disease. The disease can be minimized by adjusting the sowing dates which should be worked out separately for different agroclimatic regions. Under the Palampur conditions in India, the disease appears more severely in crop sown on 23 June as compared to that sown on 30 June or late (103). Similar observations have been made by Rewal and Bedi (284). The disease is reported to develop rapidly in plots sown on 20 June at a row to row spacing of 20 cm. The late sown crop, even at wider spacing, produced more yield due to less disease than early sown crop at closer spacing (364).

Beside, crop rotation, general field sanitation, destruction of infected crop debris and avoiding the collateral hosts in the vicinity of the main crop, help greatly in reducing the disease severity.

2.18. MYROTHECIUM LEAF SPOT :

The Disease was reported first from India in Bihar state where it caused economic losses in yield of greengram crop (382). The pathogen has been reported to cause leaf spots of blackgram also in Uttar Pradesh, India (360). The disease has been found occurring in the states of Madhya Pradesh, Rajasthan, Haryana, Himachal Pradesh, Delhi etc in India. Though, it is considered as a minor disease, it may become severe and may cause considerable losses under favourable weather conditions. The losses in yield differ greatly from variety to variety and range from 2.6% in var. T 44 to 50.6% in var. Varsha (337).

Symptoms :

The disease appears during the rainy season as small, circular, tan spots with a broad, violet to brown margin surrounded by zones of translucent areas which appear as concentric rings on which the fructification of the fungus appears as dark green sporodochia surrounded by a rim of white, hair like mycelia. The sporodochia become conical or flattened (214). The disease on mungbean is characterized by oval or circular, fading green or light brown to brown coloured spots of 0.5-1.5 cm in diameter (Fig. 2.32). The spots reveal characteristic concentric zonations impregnated with dot like bodies, the sporodochia (358). The sporodochia are formed mostly in the centre of spot and on dorsal leaf surface (Fig. 2.33). The spots may coalesce to form the bigger irregular lesions. The diseased part of the lamina may fall off leaving `shot hole' symptoms. Lesions may be formed on petioles, stems and pods also. The pathogen has been reported to cause collar rot of greengram seedlings in addition to severe leaf spotting (437).

Fig.2.32.
Myrothecium leaf spot of greem gram showing sporodochia on lower surface of leaf.

Fig.2.33.
Myrothecium leaf spot on black gram.

Causal organism :

Myrothecium roridum Tode ex Fr.

Sporodochia sessile, discoid, circular or irregular, ., 0.1 to 1.5 mm in diameter, often coalescing to form larger masses, green at first, becoming dark typically with white wooly margin composed of unicellular, contoured hyphae. *Conidiophores* erect, once or twice branched, septate, hyaline, 3-4 celled. Branches 1 or 2 celled arising from the terminal cell of the axis or immediately below the septum of the intermediate cell, each branch terminating in a whorl of phialides. The *phialides*, slenderly clavate, straight, hyaline, 10-12 x 1-2 μ forming closely packed hymenium like layer. The *conidia* cylindrical or very slightly tapering with rounded ends, 1 celled, 2-3 guttulate, hyaline at first then pale green, 5-9 X 1.5-2.5μ with an average of 8 X 2 μ (Fig. 2.34). Spore mass at first green, then jet black, viscid and finally dry and dull black (83, 363).

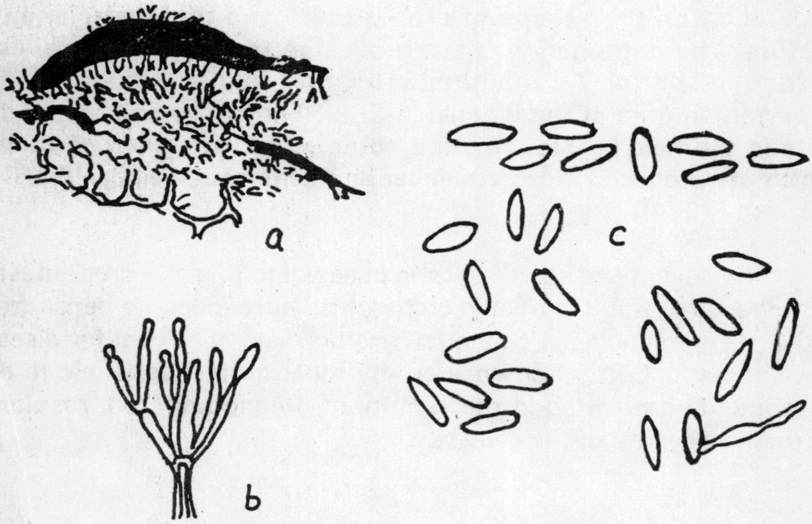

Fig.2.34.
Myrothecium roridum : a. Cross section of sporodochium, b. Conidiophores, c. Conidia.

The isolates of the pathogen from several hosts including greengram were not host specific but they differed in degree of infection on these hosts (29,243). Amongst the main crops, the pathogen can attack cowpea, soybean, cluster bean, cotton, tomato, brinjal, cucurbits, okra etc. The fungus can infect majority of the plants by providing favourable conditions for infection (243). Infact, a chemical, nicrocitin, may be responsible for such an action. It has been isolated from the culture filtrate of *M.roridum* and reported to produce the symptoms on plants as produced by the pathogen. The fungus grows as saprophyte on its host after the tissues are killed by this toxin (243).

Disease cycle and epidemiology :

The pathogen *M. roridum* has been found associated with the seeds of greengram as well as blackgram (422, 437). Thus the seed serves as the source of primary inoculum for disease development in the field. The fungus can survive on infected leaves of greengram for more than 9 months and on crop residues in the soil for 7-8 months depending on the season of crop growth and the environmental factors. The pathogen was not recovered from infected leaves if burried in soil for 2-3 months (214). Soil is however, one of the important source of the inoculum (359). Further spread of the disease in the field is through wind borne or rain transmitted conidia which are produced in large number and germinate well at 28.5°C in tap water (214).

The severity of the disease increases more in the crop infested with weeds or if it is grown in orchards or surrounded by dense trees (358). These conditions create the microclimate congineal for disease development. Cotton, cowpea or any other crop susceptible to this pathogen and growing in the vicinity of greengram crop, results in increased severity of the disease.

Control:

Chemical : Since, resistance to this disease has not been identified the use of fungicides seems the only alternative. Three sprays of tridemorph (0.05%), at 10 days interval, effectively reduce the disease incidence and check the secondary spread of the disease. Sprays of zineb or carbendazim (0.1%) have also proved effective (231). The effective control of the disease on soybean could be achieved by 2 sprays of mancozeb (0.25%) which minimized the disease severity as well as increased the yield (340). The first spray was done just after the disease appearance and the second was given after 15 days of the fist.

Cultural : Use of disease free seed and proper seed treatment with thiram 2.5 g/kg seed is of prime importance. The disease can be minimized by avoiding shady area for greengram growing; timely and proper weed control and destruction of diseased plant debris will be useful for disease control.

2.19. DRECHSLERA LEAF SPOT

A leaf spot of greengram was observed in Kanpur, India on variety T 44 (75). The disease appeared in severe form under natural conditions during August-September when continous rains and humid conditions prevailed.

Symptoms :

The symptoms appear on all the aerial parts of the plant including stems, leaves and petioles. On leaves, spots are oval in begining but gradually elongate with the midrib and axis of the leaf. At advance stages, spots coalesce to form large necrotic areas. Such leaves turn dark reddish brown, droop and wither. The lesions on stems and petioles are reddish brown, oval to oblong but coalesce frequently forming streaks.

Causal organism :

Drechslera halodes (Drechsler) Subram. & Jain

The *mycelium* is branched, septate and brown coloured. *Conidiophores,* arise singly or in groups of two, measure 60-150 μ in length and 4-7 μ in width. They are 1-5 septate, geniculate and brown in colour. Length of the segment, a typical character of this genus, very from 15-30 μ. *Conidia* are produced singly or in groups, 20-100 μ in length, 10-21 μ in width, elliptical, narrowing some what toward the broadly rounded apex, basal cell uniformly distinguished having one to three septate hilum at its base. They are brownish yellow or greenish brown to dark brown but the end of the conidial segment some time appears pale in colour (Fig. 2.35). The conidia germinate by two polar germ tubes and intermediate segment. Germ tubes are produced at thin walled region at the tip and immediately adjacent to hilum.

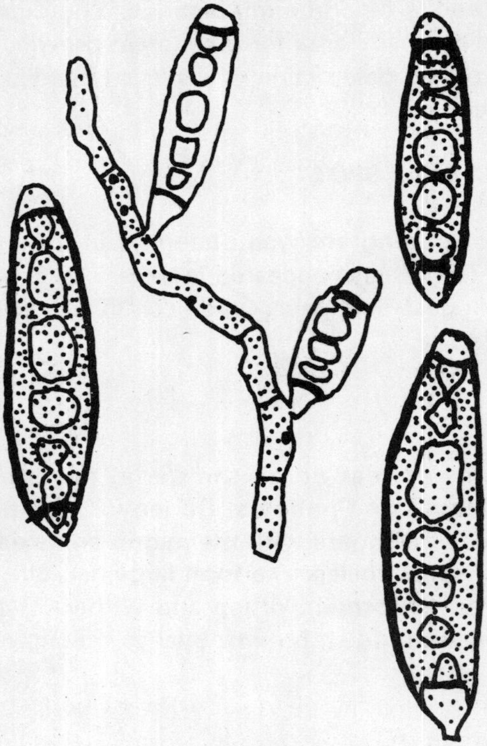

Fig.2.35.
Dreschlera halodes : Conidiophore and conidia

The pathogen is able to infect several other graminaceous hosts particularly seedlings of sugarcane in addition to greengram. Under artificial inoculation conditions it has been found pathogenic on *Lens culinaris, Pisum sativum, Vigna mungo, Coriendrum sativum, Daucus carota, Abelmoschus esculantus, Gossypium dactylon, Solanum nigrum, Cilosia cristata, Chrysenthemum* sp., *Tagets erecta* and *Euphorbia hirta*. The weed *E.hirta* is commonly found wild plant growing throughout the year and functions as the collateral host.

Host parasite interaction :

The infection of greengram plant with *D.halodes* affects its amino acid contents. It decreases lysine, valine, leucine, threonine and serine but increases the cystine. Metheonine, glycine and glutamic acid is not affected due to the infection while an unidentified additional amino acid is detected (76).

The culture filtrate of *D.halodes* is able to reduce the germination of blackgram varieties L 35-5 and T 9 and adversely affect the root and shoot elongation of seedlings. The roots are more sensitive than shoots in this regard (430).

Disease cycle and epidemiology :

The pathogen has been reported to be soil borne perpetuating through diseased plant debris from one season to other (78). Under field conditions, it can survive for more than 8 months in diseased host debris. The survival is better at low temperature in the laboratory (5$\pm$ 1°C) as compared to field conditions (18-33°C). Seed is not reported to play the role in carrying the primary inoculum of the pathogen. Plants remain free up to 2-3 leaf stage but later the symptoms appear on lower leaves near the soil level. The secondary spread of the disease is through spores produced on diseased leaves and transmitted by air.

The disease intensity has been reported to be less in crop of greengram supplied with 50 kg K_2O/ha as compared to 25 kg or no potash application. Nitrogen increases the disease while P_2O_5 has no effect (77).

Control :

In the *in vitro* tests, thiram, cuman L, ferbam, captan and aureofungin were effective. The disease was checked by spraying crop with zineb. Thiram and PCNB were effective as soil drench (80).

2.20. CORYNESPORA LEAF SPOT, TARGET SPOT :

This disease occurs on several legumes including green gram and blackgram. On blackgram the disease was first reported in India from Orissa (4). Later it was found occurring on spring crop of blackgram in Nagarjun nagar where greengram was also found susceptible to the pathogen (194).

Symptoms:

The disease appears as minute reddish brown spots scattered all over the leaf surface of blackgram. The spots gradually increase in size and assume a circular shape and vary from 2-10 mm in diameter. In severe cases, affected leaves wither (Fig. 2.36) and shed (4,194). The sporulation is mainly epiphyllous (194). Addy and Mohanty (4) reported that the conidiophores emerge singly through the epidermis on both the surfaces of leaves.

Fig.2.36.
Green gram leaf damaged by *Corynespora cassiicola.*

Causal organism :

Corynespora cassiicola (Berk & Curt.) Wei.

The fungus produces effuse, grey to brown, thinly hairy colonies with *mycelium* mostly immersed, no stroma & septate (83). The *conidiophores* are simple multiseptate, brown to dark brown and measure 100-160 X 7-10 μ (Fig.2.37). The *conidia* are terminal, borne singly, obclavate or cylindrical, olivaceous brown, multiseptate and measure 44-164 X 14-18 μ (4). The conidia produced in culture are smaller than those produced on the infected leaf (194).

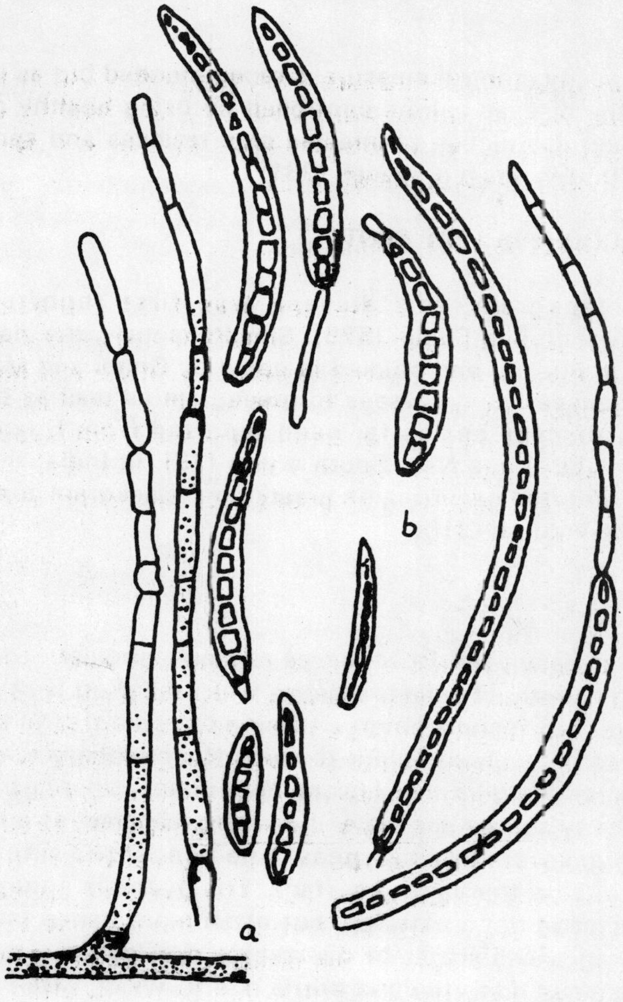

Fig.2.37.
Corynespora cassiicola: a. Conidiophores, b. Conidia

Disease cycle and epidemiology :

The pathogen has been found associated with the seed of greengram (437) and it can survive on the diseased plant debris in the field up to 2 years or on other collateral hosts The secondary spread is air borne. The disease incidence is high on crop reaching nearer to maturity. The isolate from *Salvia leucantha* is reported to infect blackgram also (17).

Control :

No specific control measure is recommended but as in case of cowpea, the disease can be controlled by using healthy, clean and treated seed; destruction of infected crop residues and spraying the crop with 0.25% zineb or maneb (231).

2.21. ASCOCHYTA LEAF SPOT :

On blackgram, this disease was first reported in the experimental crop in Sudan (375). Simultaneously, the disease was observed in Punjab and studied in detail by Ghani and Munjal (88) where it caused heavy lossess to greengram as well as blackgram crop. The disease has so far been reported from Nyasaland on greengram (361) and New South Wales (23). In India, the disease has been reported occurring on greengram, blackgram and cowpea, in Himachal Pradesh (316).

Symptoms :

The disease appears on leaves as small, circular, discoloured areas with or without definite margins when the plant is of about one month age. The lesions enlarge turning tan coloured in the centre fading in to olive green colour towards the periphery with definite dark coloured margins. The lesions have concentric rings which are more narrowly spaced and wavy in case of greengram as compared to that on blackgram and cowpeas. The concentric rings are less conspicuous on the central surface. The pycnidia appear as dark brown, pinhead dot, scattered, but often more dense towards the centre. In advanced stages of the disease development, the necrotic tissue becomes distorted and brittle due to which often shredding takes place.

Causal organism :

Ascochyta phaseolorum Sacc.

The fungus belongs to Fungi Imperfecti. The *mycelium* is septate, hyaline or subhyaline and intercellular. *Pycnidia* are mainly epiphyllous, dark, submerged to erumpent, ostiolate and measure 80 to 125 μ in diameter. *Spores* are mostly 2 celled, hyaline, oblong, smooth and measure 5.0 - 8.75 X 1.88 - 2.5 μ in size (Fig. 2.38). On PDA, the fungus forms colonies with slightly aerial, subhyaline mycelium turning dark brown to black with age. The pycnidia are produced in 7 days at 25°C (316).

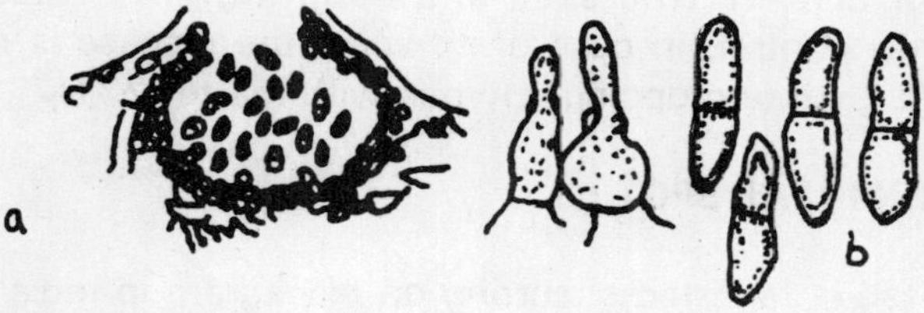

Fig.2.38.
Ascochyta phaseolorum : a. Cross section of pycnidium, b. Conidia and conidiophores (phialides).

The fungus is able to infect adzuki bean, cowpea, hyacinth bean, mothbean ano yard long bean in addition to greengram and blackgram.

The greengram has been found infected by pea isolate of *Ascochyta pinodes* (357) which is considered as the conidial stage of *Mycosphaerella pinodes*, later, reported infecting blackgram in Himachal Pradesh, India (351). Other species of *Ascochyta* reported to infect greengram are : pea isolate of *A.pisi* and *A. pinodella* (357), soybean isolate of *A.sojaecola* (175) and alfalfa isolate of *A.imperfecta* (159). The *A.pinodella* and *A.imperfecta* are considered as the synonyms of *Phoma medicaginis* var. *pinodella* and *Phoma medicaginis* respectively. The *Asçochyta imperfecta* from lucern was not found pathogenic to greengram (227).

Disease cycle and epidemiology :

Ascochyta spp. have been detected associated with seeds of greengram as well as blackgram (431). The disease spread is favoured under wet conditions and heavy soil. It can spread in pods of harvested plants stored in stooks in the field in cool moist weather and can survive in the diseased crop residues (231). The pathogen infects the plant easily under wound inoculation condition (316).

Control :

Chemital : Spray of crop with zineb @ 2.5 g/l of water has been found effective in controlling the disease satisfactorily.

Cultural : The diseased plant debris should be burnt at the end of the season. Use of disease free seed and adopting crop rotation are helpful in disease control. In case of cowpea, the disease is reduced from 53 to 30% by intercropping cowpea with maize (231).

2.22. FUSARIUM LEAF SPOT :

This disease is reported occurring on blackgram in India (36). It is not a serious disease. In addition to leaf spots the pathogen may cause seed rot and pod infection also. It reduces the protein content of the seed.

Symptoms :

The pathogen has been detected on seeds and is reported to cause seed rotting and pre-emergence mortality in blackgram as well as greengram (424, 437). It may also cause post-emergence mortality of seedlings in both the crops (228, 426, 438).

The leaf spot symptoms have been reported on blackgram (36). The disease starts as small, round to irregular, brown spots with dark coloured margins on primary leaves and later, on trifoliate leaves. Spots increase in size up to 2-3 mm in diameter and they are grey in centre with dark brown margins. Several spots may coalesce causing blighting of leaves which ultimately defoliate prematurely. In the morning hours, white fungal growth can be seen on the lower surface of the spot. Pod infections have been reported in case of green gram (424, 426).

Causal organism :

Fusarium equiseti (Corda) Sacc.

The fungus grows well on PDA aT 25°C. The *mycelium* is cottony white and produces light brown to dull pink pigmentation on medium. *Conidiophores* are branched, sparingly scattered with abundant sporulation. *Microconidia* are numerous, non septate, oval,

6-16.5 μ in length and 3-3.75 μ in width. *Macroconidia*, 3-6 septate, pale brown, typical of a twisted spindle with parabolic curvature, 12-84 μ long and 3-6 μ wide. *Chlamydospores* are round, mainly intercalary, and 6-27 μ in diameter (Fig. 2.39).

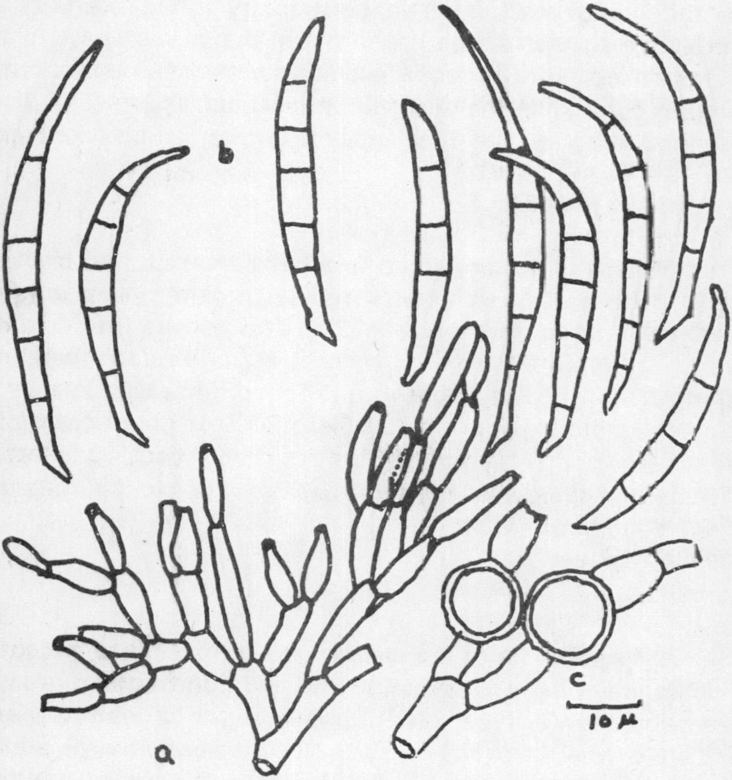

Fig.2.39.
Fusarium equiseti : a. Conidiophores, b. Macroconidia, c. chlamydospores.

Disease cycle and epidemiology :

The pathogen is mainly seed transmitted as it has been found associated with seeds of blackgram as well as greengram (422,424,426,437). The pathogen is associated internally with the seed in both the crops (424,438). It has also been isolated from the field soil (228). Thus, seed and soil may be the main sources of primary inoculum of *F. equiseti* in the field. The secondary spread on the foliage is through air and rain splash transmitted conidia. Eight day old seedlings when sprayed with the spore suspension in 0.1 % tween 20 and covered with the polythene for 24 hours, developed the disease symptoms in 7-8 days (36).

Control :

In the *in vitro* tests, bavistin, ceresan dry and agrosan GN were found effective against the pathogen but in the *in vivo* tests, dithane M 45 (mancozeb) and thiram when used as seed dressers, proved promising in controlling the seed borne inoculum (424). The disease is the minor one and hence, fungicidal spray may not be economic.

2.23. ALTERNARIA LEAF SPOT :

This disease is not important and has been reported from India only (99). The pathogen associated with greengram was first reported in Maharashtra, India (272). Later, it was found that the *Alternaria* sp. isolated from *Cyamopsis tetragonoloba* can infect greengram also (58) but under natural conditions the disease was first reported from Udaipur, India where 80% of green gram plants were found infected at an age of 65 days (99). About 45% leaf area was affected by the disease. The disease may prove injurious under favourable conditions (413).

Symptoms :

Detailed symptoms of the disease are not described except that brown lesions appear on leaves (99). In addition to green gram, the disease occurs on cowpea, lentil and pigeonpea also. Generally, spots occur at the tip or the margins. The lesions enlarge and may result in `shot hole' symptoms (231). The pathogen is reported to cause seed rot and seedling blight of greengram as well as blackgram.

Causal organism :

Alternaria alternata (Fr.) Keissler

Synonym : *Alternaria tenuis*

The identity of the pathogen isolated from greengram (99) was confirmed by Dr. M.B.Ellis, CMI, England. The pathogen (Fig. 2.40) has specific requirements for its proper growth in regard of sources of carbon, nitrogen and the growth regulators (100). *Alternaria tenuis* is reported to produce a toxin which causes albinism in green gram seedling (441). The culture filtrate of the pathogen isolated from seeds of greengram, reduces the germination of blackgram and inhibits root and shoot elongation greatly (432). A toxin has been purified from the culture of *A.alternata* which disrupts the normal etioplast and chloroplast development by the primary leaves of greengram (73). It occurs through a mechanism unrelated to photophosphorylation.

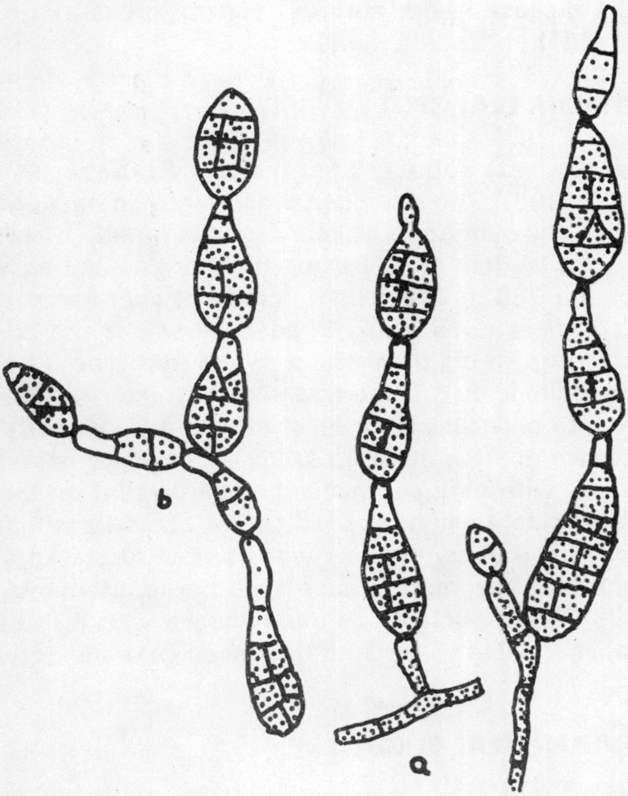

Fig.2.40.
Alternaria alternata : a. Conidiophores bearing *chains* of conidia,
b. Branched chain of conidia.

Disease cycle and epidemiology :

The pathogen has been reported to be seed transmitted in green gram and blackgram (440). Further transmission of the disease is through the wind and rains (231). The disease severity is influenced by changing the sowing time (279). Greengram plants when spray inoculate at age of 65 days and incubated at the temperature of 28°C under saturated conditions of humidity, developed symptoms in 3 - 4 days.

Control :

Since the disease is not serious, control measures are not recommended (231).

2.24. ASCOTRICHA LEAF SPOT :

The disease was reported from Uttar Pradesh, India on blackgram (302, 303). The pathogen was identified as *Ascotricha chartarum* Berk. The symptoms appear as small round, 1-2mm size, grey spots on the leaflets mostly along the margins and advancing towards the midrib (302). The central portion of spot becomes light grey at first and later turns black. In advance stages the adjoining spots coalesce to form big patches up to 4-8 mm. The lesions are noticed on the entire leaflet. Some time they become papery and the dead tissue fall to give the `shot hole' effect. A collar rot disease was also prevalent and the diseased seedlings could be easily pulled out from the soil with dark perithecial bodies present on the basal portion of stem up to a height of 4-10 cm (302). The pathogen has been isolated from leaflets, stem as well as seed of blackgram and greengram and is suspected to cause seed borne infections in the field (300, 303). The association of the pathogen with the seed was extra embryal as it could be isolated from seed coat and cotyledons only.

2.25. CURVULARIA LEAF BLIGHT :

A leaf blightof greengram caused by *Curvularia eragrostidis* (Henn.) Meyer (Fig. 2.41) was reported from Assam, India (33). The disease appeared in later part of October. Initially, the disease manifests as small, irregular, brown, spots at the apices usually about 1-4 mm long which increase in size, become necrotic and

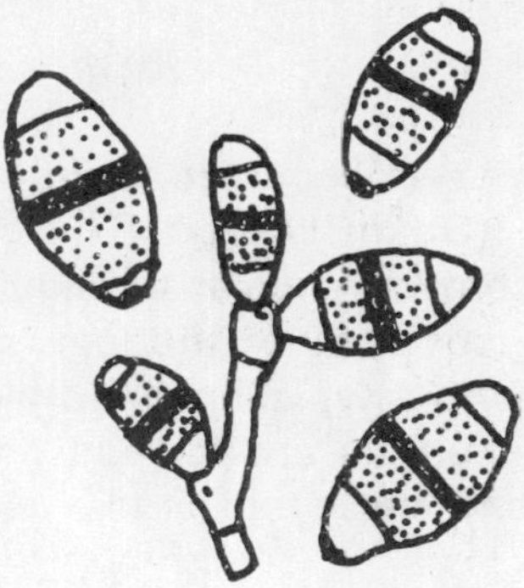

Fig.2.41.
Curvularia eragrostidis : Conidiophore and conidia

covered with copious growth of fructifications.

Under artificial inoculations, symptoms developed after 3-4 days on wounded and 8-10 days on un-wounded leaves. Out of 11 varieties of greengram infected in the field, minimum disease incidence was on K 851 (30%) as compared to 75-80% on Madira and ML 24. Incidence on T 44 was 61 % and the pathogen attacked members of *Arachis* and pine apple. *Curvularia* sp. have frequently been reported on seeds of greengram as well as blackgram so may be the case with *C. eragrostidis* also. No control is recommended but seed treatment will help in checking the seed borne pathogen.

2.26. DACTULIOPHORA LEAF SPOT :

A leaf spot of greengram caused by *Dactuliophora tarrii* Leakey was reported from Wadoora in Kashmir, India by Koul (174). The disease is reported to appear in the form of zonate spots of variable shapes and 1-3 cm in size. These spots usually appear on the upper surface of the leaf as violet pin head dots which increase in size and develop in to distinctly zonate spots with alternating whitish and tan coloured bands. The spots often remain distinct from one another and the green colour of the leaf around the spot becomes a little paler. The lower surface is densely covered with the sclerotia of the fungus, particularly on the paler part, and the zonation is less distinct on this side of the leaf. A leaf may show numerous spots if the disease is well advanced. It may result in premature defoliation. The pathogen has been mentioned to be a new Indian record.

2.27. SEPTORIA LEAF SPOT :

This disease was also reported from India (273) caused by *Septoria leguminum* (Desm.) Sacc. The symptoms appear as numerous, small, pale brown, circular to angular spots of 2-8 mm in diameter on both the surfaces of the leaf. Later, colour of spots changes to brick red with a typical pale purple central zone. Minute black pycnidia of the fungus are formed on spots. The affected leaves shrivel, dry up and finally exhibit blighted appearance.

2.28. PHOMA LEAF SPOT :

A leaf spot disease of greengram caused by *Phoma medicaginis* var. *pinodella* (Jones) Boerema *et al.* was reported from Jabalpur, India by Rajak and Rai (264). They reported it to be a new host record of the fungus, though, Sprague earlier (357) reported the greengram to be infectd by the pea isolate of *Ascochyta pinodella* which is now considered as the synonym to *P.medicaginis* var. *pinodella*. The pathogen is characterized by : *colonies* flat, grey with abundant pycnidia on oat agar produced in radial rows; *pycnidia*, globose to subglobose, parenchymatous, ostiolate, 75-148 μ; *pycnidiospores*, hyaline, 1 celled, ovoid to ellipsoid, 4-7.4 X 2-3.5 μ in size and *chlamydospores* in chains, spherical, some times septate.

Another species, *Phoma exigua* Desm., which mainly deteriorates the seed of green gram, has been found causing leaf spot of blackgram in India (338). It is a minor disease and not reported from any other place.

2.29. PHYLLOSTICTA LEAF SPOT :

Taxonomically *Phyllosticta* is not different from the genus *Phoma* except the conventional base of different organs of plant host attacked by them. Most of the species of *Phyllosticta* have been transferred to other genera. *Phyllosticta phaseolorum* was reported to cause serious leaf blight of greengram by Ghani and Munjal who discussed the taxonomy and pathogenicity of *P.phaseolorum* and *Ascochyta phaseolorum*. It was later reported from New South Wales (22) occurring on greengram. Simultaneously another species of *Phyllosticta* was reported occurring on *P. mungo* var. *radiatus* from India as *Phyllosticta phaseolina*. (271).

2.30. HETEROSPORIUM AND CHAETOSEPTORIA LEAF SPOTS :

Leach (185) reported dark brown, angular interveinal spots consisting of a velvety fungal growth on the under surface of greengram leaves from Nyasaland. The fungus resembled *Heterosporium* sp.

Chaetoseptoria wellmanii is reported to infect several leguminous hosts including *Phaseolus mungo* causing leaf spot disease in Central America (213).

2.31. MUNGBEAN SCAB :

The first record of mungbean scab was from Kenya (221) where *Elsinoe phaseoli* was reported causing serious infections of leaves, stem and pods of greengram. The disease appeared there seriously again in 1955 (19). Later, the disease was reported from Indonesia (147) where it attacked all the greengram varieties seriously. The diseases is not reported from any other country so far but it may cause yield losses up to 60% particularly under wet weather conditions (211). Blackgram is not attacked by the scab pathogen.

Symptoms :

The disease appears on leaves, petioles, stem and pods. In severely infected plants, young leaves curl and plant remain stunted. Detailed symptoms have been described by Mukelar (210). The lesions on leaves are brown to reddish brown, circular and 1-2 mm in size which gradually enlarge up to 5 mm and occasionally become angular in shape. Often the tissues around the lesion turn yellow. The centres of the old lesions become grey or greyish white and drop out to give `shot hole' appearance. Lesions formed on or along the midrib are more conspicuous on lower side of leaf. Severely infected young leaves are curled and distorted.

On the stem, the lesions are round or oval, 3-5 mm in diameter and grey to greyish white in the centre surrounded by reddish brown border. Some times they coalesce forming 1 cm or more long spots which are slightly raised showing typical scab symptoms. Lesions formed on immature pods are more conspicuous. Individual lesions are sunken, oval elliptical, semicircular or some times irregular in shape. Their size may range from mere flecks to 5-8 mm and they are dark brown to reddish brown in colour with grey centres. With the maturity of pods they become raised and lighter in colour.

Causal organism :

Elsinoe iwatae Kajiwara et. Mukelar sp. n.

Elsinoe phaseoli was reported infecting mungbean in Kenya (221) but details were not given. It is reported to cause scab of common bean, cowpea and yard long bean (231). The cause of mungbean scab was studied in detail in Indonesia by Kajiwara and Mukelar (148,149,210) who named it *E. iwatae.*

Acervuli develop in the central region of recently formed lesions and in the marginal zone of younger lesions. They are disc or cushion shaped. *Conidiophores* are very short and closely compacted producing abudant *conidia* under favourable conditions. They are scarcely formed on the old lesions and are hyaline, one celled, oblong or elliptical and 5-6.5 X 2-3 μ in size. The *asci* are formed in stroma under the host epidermis singly or in groups, some times with their wall in contact. Perithecia are absent. Asci are hyaline, ovoid, globose to ellipsoid, 25-27 X 17.5-22.5 μ in size. In water, they rupture releasing 8 *ascospores* which are obovate or elliptical, hyaline, 2-3 septate and 11-14 X 5-7.5 μ in size.

It grows very slowly on PDA reaching 1 cm after 2 months at 22-26°C. Colonies are yellowish in the beginning but later turning reddish brown with wrinkled surface. Conidia are not formed on the PDA. Morphologically, the mungbean scab fungus is different from *E.phaseoli* but it is some what similar to *E.canavaliae, E.dolichi* and *E.calopognii*. However, it differs in pathogenicity.

The host range of the mungbean scab pathogen is not very wide. It is pathogenic mainly to greengram but can infect *Phaseolus angularis, Dolichos lablab* and *Pisum sativum*. It is unable to infect several other legumes including *Glycine max, Phaseolus lunatus* and *Canavalia gladiata*. On the basis of the morphology and the pathogenicity tests, Mukelar (210) mentioned possibility of the pathogen to be considered as the forma specialis of *E.canavaliae*. It is however, necessary that further investigations on the pathogenicity of each, *Elsinoe* and *Sphaceloma* spp. from the leguminous plants, be clearly designated before unification. The pathogen does not infect sword bean and lima bean and thus differs from *E.canavaliae* and *E.phaseoli*.

Disease cycle and epidemiology :

No work on this aspect of the disease has been done but *E.phaseoli* is reported to be seed borne and its air dispersal is also suspected. It can survive on crop debris (231) and becomes severe under wet weather conditions.

Host resistance :

Greengram varieties and accessions were screened in Indonesia under natural infection. The resistance was observed in Black matse, Kacang hijau, Black grain variety Gwalior 2, Kacang hitam, *Phaseolus mungo* No. CPI 12106 and 3335, Boc E. Kojae, Kassa, Minoe 505 and Kacang hijau I (210). Resistance in greengram varieties against scab has been identified in Taiwan also (231).

Control :

The fungicides,topsin M 70 WP (thiophanate methyl), bavistin 50 WP (carbendazim) and benlate 50 WP (benomyl) have been found effective in controlling the disease. Two sprays of any of the fungicide at 0.19 kg a.i./ha control the disease effectively increasing the yield significantly over the control (211). First spray is recommended at 20 days and the second at 40 days after sowing. The increased yield in treated plots reaches about 2.5-2.6 times that of the control (210).

Destruction of diseased crop debris may help in reducing the disease severity.

POD ROTS

Several fungi have been reported to cause rotting of pods of greengram as well as blackgram resulting in severe yield losses particularly under wet weather conditions.

2.32. DIPLODIA POD ROT :

This disease was first observed on greengram in Philippines in the year 1976 (129, 131). The disease may cause serious losses as entire pods blacken and rot before maturity.

Symptoms :

Usually green pods are infected in which the discolouration starts from the stem end and progresses towards the tip of the pod. At the initial stage, the infected pods are soft and soggy but later they dry and become hard. If the infection is at pod filling to fully developed stage of the seed, the seeds become dark and shrivelled. Under artificial inoculation of unfilled young pods, the symptoms appear 2 days after inoculation in the form of soft rotting. The infection spreads very fast so that the entire pod rots and becomes dark brown to black within 5 days of inoculation (129, 131).

Causal organism :

Diplodia natalensis (= *Botryodiplodia theobromae*)

The mungbean isolate of the pathogen has been described in Philippines (131). The *pycnidia* are globose or subglobular, black, occur singly, immersed in host tissues, and erumpent. *Conidia* are one celled and subhyaline, becoming dark brown, two celled and striated with the age. They are ovoid or ellipsoid, non constricted and are borne on simple *conidiophores*. The pycnidia measure 285.6-938.4 μ in diameter while conidia are 16-38 x 11-17 μ with an average of 28 x 14 μ in size.

The pathogen can infect *Phaseolus vulgaris, Vigna sinensis* and *V.sesquipedalis* v. *sinensis* causing soft rot.

Disease cycle and epidemiology :

No work has been done on this aspect but the pathogen is probably seed-borne in nature. The fungus *Botryodiplodia theobromae* has been detected in the seeds of greengram as well as blackgram in Madhya Pradesh, India (431).

Control :

No control is known but avoiding the wet weather conditions during the pod formation stage by shifting the sowing dates may help in reducing disease incidence. In Philippines, 10 greengram varieties were tested under artificial inoculation conditions of which 3 were moderately resistant (131).

2.33. CHOANEPHORA POD ROT :

It is a serious disease of greengram, blackgram, mothbean and cowpea reported from Rajasthan, India (196).

Symptoms :

The disease symptoms appear as water-soaked areas on green pods which later turn brownish black. Soon after, a luxurient, white fungal growth develops on the infected part and constitutes of the conidial and sporangial fructifications. The fructifications look as minute, black, pin head like structures. The infected pods rot completely within few days of infection.

Causal organism :

Choanephora cucurbitarum (Berk & Rav.) Thaxter.

The fungus produces abundant sporangia and conidia in culture. The *sporangia* are terminal and pendant on the curved ends of erect *sporangiophores.* Each sporangium possesses a definite columella and measures 38.5-152.0 μ in diameter. *Sporangiospores* are light brown, ovoid, smooth and provided with a tuft of hair like appendages at both the ends. They measure 14.5-23.5 x 8.0-14.0 μ in size. The *conidiophores* are erect ending in a capitate vesicle from which short secondary vesicles are produced. These vesicles, at the maturity are covered with short sterigmata bearing conidia. The *conidia* are light brown to reddish brown, ovoid, longitudinally striate and often provided at the base with a short hyaline appendage. They measure 13.5-20.5 x 8.0-12.5 μ in size (Fig. 2.42). The *zygospores* are not observed on these crops.

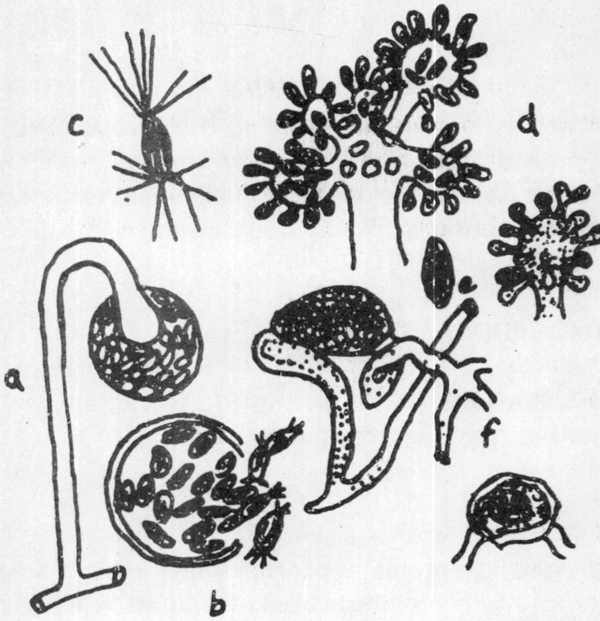

Fig.2.42.
Choanephora cucurbitarum : a. Sporangium with sporangiophore, b. Dehiscing sporangium, c. Sporangiospore d. Heads of conidiophore with round headed capituli e. Conidium, f. Young and matured zygospores.

Disease cycle and epidemiology :

The *Choanephora* sp. has been found asociated with seeds of greengram and is expected to be seed-borne (437). The disease becomes severe during rainy season when the high humidity persists. Injury on pods facilitate the infection and fast development of the disease.

Control :

Not known.

2.34. PYTHIUM POD ROT :

This disease was reported occurring on greengram, blackgram and cowpea in Rajasthan, India (195).

Symptoms :

The disease is characterized by typical cottony growth of the fungus on pods. Under cool and humid environment, the disease manifests as large water-soaked areas on green pods. In the beginning water-soaked lesions get covered with a thin layer of cottony growth which later becomes thicker to form a cottony velvet. In some cases a dirty fluid may ooze out of the infected pods.

Causal organism :

Pythium debaryanum Hesse

The fungus is characterized by terminal or intercalary, spherical to oval *sporangia* which measure 15-26 μ in size and germinate by germ tubes or releasing *zoospores* (453). *Oogonia* are terminal or intercalary, smooth, usually spherical, 15-28 μ in size. *Antheridia*, monoclinous or diclinous, one to six per oogonium. *Oospores* are smooth, aplerotic, 12-20 μ and germinate by germ tube which at low temperature give rise vesicle in which zoospores are formed.

2.35. FUSARIUM POD ROT :

A pod rot of greengram and blackgram caused by *Fusarium solani* (Fig. 2.43) is reported occurring in India (45). the disease starts as gradual drying of pods from the distal ends to basal ends. The lesions are light brown to greyish brown without definite margins. They enlarge covering whole pod in 7-10 days. Later, pods shrivel and dry. Severely infected pods do not contain seeds but show fungal growth inside. In partially infected pods, no effect is observed on seed except reduction in their size. In case of early infection, the development of the pod is checked. The surface of such pod is covered with white mycelial growth under humid conditions.

2.36. CURVULARIA POD ROT :

This pod rot is caused by *Curvularia lunata* in greengram as well as blackgram (45). The disease starts as pale yellow, circular spot on pod which later becomes brown to black in colour. Heavily infected pods contain cracked seeds with black scars.

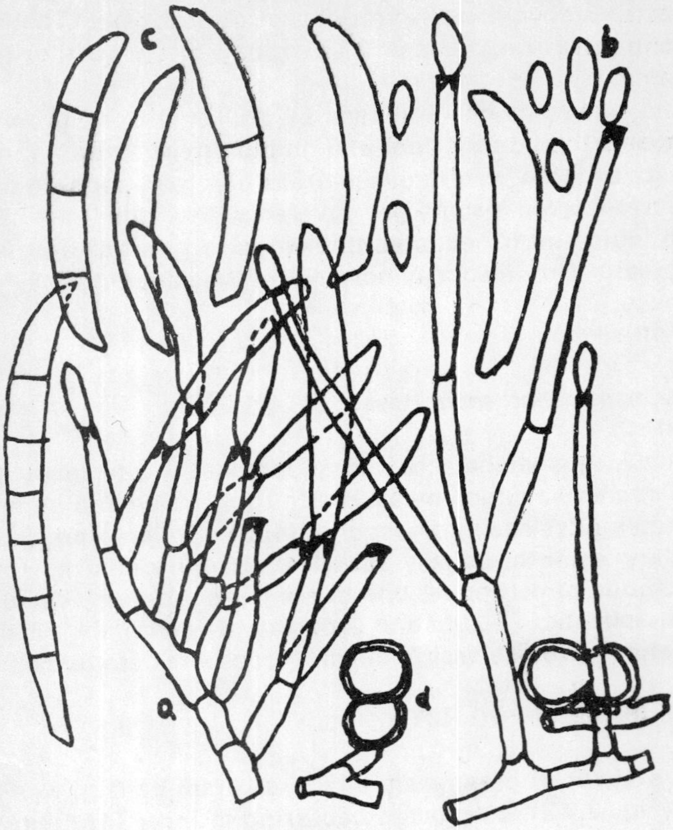

Fig.2.43.
Fusarium solani : a. Conidiophores, b. Microconidia, c. Macroconidia, d. Chlamydospores.

2.37. SPROUT ROT :

The mungbean is used as sprouts in several countries. In U.S.A., the mungbean sprout rot has been reported to be destructive (64).

Symptoms :

The disease appears as small, pink to orange spots on cotyledons or hypocotyl. Under the moist conditions required for sprouting, the sprouts become slimy and grief. When the inoculated seeds are placed in the sprouting chambers, the characterstic lesions develop after 3 days and the sprouts rot after 5 days. The highest spore concentration induces 94% decay of sprouts but lower concentration reduces the decay.

Causal organism :

Cylindrocephalum sp.

The colony of the fungus is salmon-orange and zonate with white and orange bands. The outer zone takes speckled grey appearance. The *conidiophores* are 20-30 μ long and the conidiogenous area has small indistinct collarettes. *Conidia* are hyaline, ellipsoidal, 11-14 x 3.8-5.2 μ, aggregating in clusters in the conidiogenous region.

Control :

The initial source of the inoculum seems on seed. No symptoms are produced on plants grown from the inoculated seeds. Hence it has been advised to periodically disinfect the sprouting containers in 1% sodium hypochlorite for 12 h to reduce the build up of the inoculum.

2.38. YEAST SPOT OF MUNG BEAN :

In South Africa, a cotton ball disease was observed with which *Nematospora gossypii* and *Nematospora coryli* were found associated (438). They were found associated with seeds of several bean varieties including mungbean. *N. coryli* is reported to cause yeast spot of seeds of soybean and mungbean in United States (253, 254). It causes sunken, pale-cream coloured spots on seeds which in severe cases remain small and shrivelled. The infection mostly follows puncture of pods by southern green strink bug. The seeds infected early may not mature or they may remain small and shrivelled. On developing seeds, infected tissue is slightly depressed. Seed shrivelling due to the yeast spot and strink bug injury can easily be differentiated from the other causes by locating the small discoloured puncture in the former case.

REFERENCES

1. Abbaiah, K. 1989. Weather factors in relation to epidemics of powdery mildew on urdbean. *National Symposium on New Frontiers of Pulses Research & Development,* Kanpur, 115-116 pp.

2. Abu kassim, A.B. 1981. Pest and disease problems of mungbean in West Malaysia. *Malaysian Agricultural Journal* 53 : 29-33.

3. Abu Kassim, A.B., Abdul Rahman, H.B. and Zainal A.B. 1977. Mungbean in West Malaysia. *First International Mungbean Symposium, AVRDC, Taiwan,* 15-17 pp.

4. Addy, S.K. and Mohanty N.N. 1960. Leaf spot disease of blackgram (*Phaseolus mungo* L.). *Proceedings of 47th Indian Science Congress Association,* 330-331 pp.

5. Agrawal, K.C. and Philip R. 1983. Economic control of powdery mildew of urid *(Phaseolus mungo),* by wetsulf. *Indian Phytopathology* 36 : 359-361.

6. Agrawal, S.C. 1989. Status of foliar fungal diseases of mungbean and urdbean and their control in Madhya Pradesh. *National Symposium on New Frontiers of Pulses Research & Development, Kanpur,* 114-115 pp.

7. Agrawal, S.C. and Nema Sushma. 1988. Resistance in mungbean and urdbean against powdery mildew. *Indian Journal of Pulses Research* 2:76-77.

8. Agrawal, S.C. and Nema Sushma. 1989. Effect of carbendazim on Macrophomina leaf blight of blackgram and greengram. *Indian Journal of Plant Protection* 17 : 147-149.

9. Agrawal, S.C. and Lilhore D.R. (under publication). Transmission of *Macrophomina phaseolina* causing leaf blight in blackgram. *Indian Journal of Mycology and Plant Pathology.*

10. Ahmad, Q.A. 1949. Varietal resistance and susceptibility of *Phaseolus radiatus* towards *Cercospora. Science & Culture* 14 : 436.

11. Ahmad, Q. 1985. fungicidal control of Cercospora leaf spot of mungbean (*Vigna radiata). Indian Phytopatholgy* 38 : 418 -422.

12. Ahmad, Q. 1987. Residual fungitoxicity and spray time interval of a few fungicides against *Cercospora canescens. Indian Phytopathology* 40 : 542-543.

3. Alam, S.S., Qureshi S.H. and Bashir M. 1985. A report on web blight of mungbean in Pakistan. *Pakistan Journal of Botany* 17 : 165.

4. Amin, K.S. and Singh R.A. 1987. Diseases of mung, urid and pea and their control. In "Plant Protection in Field Crops" (M.Veer Bhadra Rao and S. Sithananthan eds.), Plant Protection Association Of India, 203-217 pp.

5. Amin, K.S. Singh R.A., Vishwa Dhar and Gurha S.N. 1988. Changing fungal disease situation in pulse crops. *proceedings National Seminar on Changing Pest Situation in the Current Agriculture Scenario of India* 68-69 pp.

6. Amu Singh, S. 1984. *Ampelomyces quisqualis* Ces on blackgram. *FAO Plant Protection Bulletin* 32 : 144.

7. Anahosur, K.H., Fazalnoor K. and Narayanaswamy B.C. 1972. A new leaf spot disease of *Salvia leucantha* Cav. from India. *Sydowia* 25 : 143 - 146.

8. Anderson, T.R. 1985. Root rot and wilt of mungbean in Ontario. *Canadian Plant Disease Survey* 65 : 3-5.

9. Anonymous. 1957. Annual Report 1955. Department of Agriculture Kenya, II : 237.

10. Anonymous. 1960. Work of IRCT (1957-1960). *Cot. et. Fibr. Trop.* 14(2) : 83-327.

11. Anonymous. 1964. Division of Plant Industry, Science branch, Plant Pathology Section Report (1963-64), 24-25 pp.

12. Anonymous. 1965. Plant disease survey for the year ending 30th June 1964. 34th Annual Report N.S.W. Department of Agriculture, Division of Science Services, Biology branch, 38 pp.

23. Anonymous. 1965. New Plant Diseases. *Agriculture Gazet New Sou Wales* 76 (1) : 38-41.

24. Arjunan, G., Vidhyasekaran, P. and Kolandaisamy S. 1976. How combat diseases of greengram. *Farmer & Parliament* 11(2) : 17-18.

25. Arora, Y.K. and Bajaj, K.L. 1977. Effect of etherol on the phenolics mung (*Phaseolus aureus*) infected by *Rhizoctonia solani. Ac Phytopathologica Academiae Scientiarum Hungaricae* 12 : 269-271.

26 Arora, Y.K. and Bajaj K.L. 1978. Phenolic changes in mung (*Phaseol aureus*) infected by *Rhizoctonia solani. Acta Phytopathologi Academiae Scientiarum Hungariciae* 13 : 337-341.

27. Arora, Y.K. and Bajaj, K.L. 1978. effect of vitavax on the polypheno constituents of mung infected by *Rhizoctonia solani. Plant Biochemic Journal* 5 : 97-100.

28. Arora, Y.K. and Bajaj K.L. 1985. Peroxidase and polyphenol oxida associated with induced resistance of mungbean to *Rhizoctonia sola* Kuhn. *Phytopathologische Zeitschrift* 114 : 325-331.

29. Arya, H.C. 1959. On the occurrence of physiological strains *Myrothecium roridum. Indian Phytopathology* 12 : 164-167.

30. Ashraful Islam, M. 1977. Mungbean Cultivation in Bangla desh. *Fir International Mungbean Symposium, AVRDC, Taiwan* 21-23 pp.

31. AVRDC. 1982. Asian Vegetable Research and Development Centr Centerpoint Vol. 3 No. 1, Shanhua, Taiwan, Republic of China.

32. Aycock, R. 1966. Stem rot and other diseases caused by *Sclerotiu rolfsii. North Carolina Agricultural Experiment Station Technical Bulle No.* 174, 102 pp.

33. Barman, B., Hyder A. and Roy A.K. 1976. Leaf blight of greengra caused by *Curvularia eragrostidis. Current Science* 45 : 468.

34. Baruah, P. and Lal S. 1981. Host range of *Rhizoctonia solani* f.s *sasakii,* the incitant of banded sclerotial disease of maize. *Indi Phytopathology* 34 : 494-496.

35. Bshir, M., Alam S.S., Qureshi S.H. and Malik B.A. 1985. Control mungbean anthracnose by foliar fungicides. *Pakistan Journal Agricultural Research* 6 : 173-175.

36. Beniwal, S.P.S. and Tripathi H.S. 1979. Leaf spot of urdbean caus by *Fusarium equiseti* (Corda) Sacc. *Tropical Grain Legume Bulletin* 1 15-16.

37. Beniwal, S.P.S., Saxena G.C. and Tripathi H.S. 1983. Natural occurrence of anthracnose of mungbean caused by *Colletotrichum capsici*. *Indian Journal of Mycology and Plant Pathology*. 13 : 356-357.

38. Beyries, A. 1969. Effectiveness of some systemic fungicides against *Corticium solani* by seed treatment of bean and radish. *Phytiat-Phytopharm*. 18 : 107-116.

39. Beyries, A. and Massiaen C.M. 1969. Virulence and specificity of some strains of *Corticium solani* in relation to market crops. *Annls. Phytopathology* 1: 37-54.

40. Bhardwaj, C.L., Paul Y.S. and Singh B.M. 1985. Some new leguminous hosts of powdery mildew. *Indian Journal of Mycology and Plant Pathology* 15 : 292.

41. Bhardwaj. C.L. ans Singh B.M. 1984. Host range of *Oidium* state of *Erysiphe pisi* from peas on some leguminous hosts in Kangra valley of Himachal Pradesh. *Indian Phytopathology* 37 : 732-733.

42. Bhardwaj. C.L. and Singh B.M. 1986. Strain variation in *Colletotrichum dematium* f. sp. *truncatum* from four leguminous hosts. *Indian Journal of Mycology and Plant Pathology* 16 : 139-141.

43. Bhargava, K.S. and Gupta S.C. 1957. Market diseases of fruits and vegetables in Kumaon I. Cottony leak of beans. *Indian Phytopathology* 18 : 48-49.

44. Bhargava, S.N. 1965. Studies on the charcoal rot of potato. *Phytopathologische Zeitschrift* 53 : 35-44.

45. Bhargava, S.N. and Shukla C.N. 1978. Pod rot of *Phaseolus mungo* and *Phaseolus aureus*. *Acta Botanica Indica* 6 (Suppl.) : 192 - 193.

46. Bhargava, S.N., Shukla D.N. and Singh N. 1979. correlation between the number of root nodules and incidence of root rot and wilt diseases of pulses. *Proceedings National Academy of Sciences India* B 49 : 56.

47. Bhaskara Rao, P. and Mallaiah K.V. 1985. Factors affecting germination of *Cercospora canescens* pathogenic to blackgram. *Indian Phytopathology* 38 : 559-560.

48. Bhaskara Rao, P. and Mallaiah K.V. 1987. Viability of *Cercospora canescens* conidia under simulated air borne condition. *Current Science* 56 : 1070 - 1071.

49. Bhaskara Rao, P. and Mallaiah K.V. 1988. Factors affecting sporulation of *Cercospora canescens* pathogenic to blackgram. *Indian Phytopathology* 41 : 280.

50. Bhaskaran, R. and Padmanabhan P. 1978. Changes in total phenols and sugars in mung and urdbean due to infection by powdery mildew fungus. *Food Farming and Agriculture* 10 (6) : 232 - 233.

51. Bhate, S.B., Utikar P.G. and Shinde P.A. 1985. Chemical control of damping off of mung. *Journal of Maharashtra Agriculture University* 10:

52. Bhate, S.B. Utikar P.G. and Shinde P.A. 1985. Reaction of mungbean varieties to damping off and root rot. *Journal of Maharashtra Agriculture University* 10 : 230.

53. Bhikane, N.S. and Mukadam D.S. 1989. Anti callus property in the culture filtrate of three seed borne fungi in greengram. *Indian Phytopathology* 42 : 349.

54. Briton-Jones, H.R. and Baker R.E.D. 1934. Notes on some other fungous diseases in Trinidad, 1933, *Tropical Agriculture (Trinidad)* 11 : 67-68.

55. Butler, E.J. 1918. "Fungi and Diseases in Plants". Thacker Spink & Co. Calcutta, 547 pp.

56. Camporata, P. 1982. Research on the ecology of parasitic fungi in the soil. XVII. Measurement of the infectious potential of soils and substrates infested by *Rhizoctonia solani* Kuhn, causal agent of seedling damping off. *Agronomie* 2 : 437-442.

57. Chamberlain, D.W. 1948. Rotation only control of stem rot. *Soybean Digest* 8 (10) : 18.

58. Chand, J.N. and Verma D.S. 1968. Occurrence of new *Alternaria* leaf spot of cluster bean in India. *Plant Disease Reporter* 52 : 145-147.

59. Cheralu, C., Narasimhan V. and Satyanarayana A. 1989. Inheritance of resistance to angular black leaf spot disease in mungbean. *National Symposium on New Frontiers in Pulses Research and Development, Kanpur*, 35 p.

60. Chohan, J.S. 1968. Disease problems in pulse crops. *Proceedings 2nd Annual Workshop Conference Pulses, IARI, New Delhi*.

61. Chohan, J.S. and Kaur J. 1976. Cultural studies on *Rhizoctonia batataticola*, casual agent of root rot of sunflower. *Indian Journal of Mycology and Plant Pathology* 6 : 140-144.

62. Chona, B.L. and Munjal R.L. 1956. Notes on miscellaneous Indian fungi. III. *Indian Phytopathology* 9: 53-56.

63. Chupp, C. 1953. "A Monograph of the Fungus Genus *Cercospora*". Ithaka New York, 667 pp.

64. Cody, Y.S. and Maloy O.C. 1984. *Cylindrocephalum* rot of mungbean sprout. *Plant Disease* 68 : 304-305.

65. Deshkar, M.V., Khare M.N. and Joshi L.K. 1973. Paper towel method, a rapid technique for testing susceptibility to *Rhizoctonia bataticola* (Taub.) Butler. *JNKVV Research Journal* 7 : 300-301.

66. Deshkar, M.V., Khare M.N. and Singh L. 1974. Evaluation of varieties of mung (*Phaseolus aureus* Roxb.) for their resistance to *Rhizoctonia bataticola* (Taub.) Butler by paper towel method. *JNKVV Research Journal* 8 : 60-62.

67. Devi, T.P. and Samy P.N. 1986. Culturing of powdery mildew on blackgram and greengram callus tissues. *Madras Agricultural Journal* 73 : 474-475.

68. Dhingra, O.D. 1971. The relationship of *Rhizoctonia bataticola* to the fungal flora in rhizosphere of urid (*Phaseolus mungo* L.). M.Sc. Ag. Thesis, JNKVV, Jabalpur, M.P., India, 88 pp.

69. Dhingra, O.D. and Khare M.N. 1973. Biological control of *Rhizoctonia bataticola* on urid bean. *Phytopathologische Zeitschrift* 76 : 23-29.

70. Dixon, G.R. 1981. "Vegetable Crop Diseases". MACMILLAN Publishers Ltd., London, 404 pp.

71. Dubey, S.C. and Dwivedi R.P. 1989. Mode of penetration by *Thanatephorus cucumeris* (= *Rhizoctonia solani*) in some leguminous hosts. *Indian Phytopathology* 42 : 317.

72. Dublish, P.K. and Pande P.C. 1976. Effect of culture filtrates of some rhizosphere fungi on seed germination and seedling growth of *Phaseolus mungo* L., *Luffa cylindrica* M. Roem and *Cucumis sativus* L. *Bangladesh Journal of Botany* 5 : 39-42.

73. Duke, S.O., Wickliff J.L., Vaughn K.C. and Paul R.N. 1982. Tentoxin does not cause chlorosis in greening mungbean leaves by inhibiting phosphorylation. *Physiologia Plantarum* 56 : 387-398.

74. Dunleavy, J. 1967. Red clover infected by *Cephalosporium gregatum*. *Phytopathology* 57 : 810.

75. Dwivedi, K. and Shukla P. 1979. A new leaf spot disease of mung caused by *Drechslera halodes. Indian Journal of Mvcology and Plant Pathology* 9 : 281.

76. Dwivedi, K. and Shukla P. 1980. Amino acid changes in mung foliage infested with *Drechslera halodes. Indian Phytopathology* 10 : 192-193.

77. Dwivedi. K. and Shukla P. 1981. Effect of interaction of nitrogen, phosphorus and potash on Drechslera leaf spot of mung. *Indian Phytopathology* 34 : 200-202.

78. Dwivedi, K. and Shukla P. 1983. Perpetuation of *Drechslera halodes* causing leaf spot disease of mungbean. *Indian Phytopathology* 36 : 427-430.

79. Dwivedi, K. and Shukla P. 1983. Some new hosts of *Drechslera halodes. Indian Journal of Mycology and Plant Pathology* 13 : 202.

80. Dwivedi, K. and Shukla P. 1983. Evaluation of some fungicides and antibiotics against *Drechslera halodes* causing leaf spot of mung. *Pesticides* 17(4) : 21-22.

81. Dwivedi, R.P. and Saksena H.K. 1974. Occurrence of web blight by *Thanatephorus cucumeris* on mungbean. *Indian Journal of Farm Science* 2 : 100.

82. Elazegni, F.A. and Mew T.W. 1983. Comparative effects of fungicides on pre-emergence damping off of green legumes caused by *Rhizoctonia solani, Sclerotium rolfsii* and *Pythium debaryanum. Tropical Grain Legume Bulletin* No. 27 : 2-7.

83. Ellis, M.B. 1971. "Dematiaceous Hyphomycetes". CMI, England, 608 pp.

84. Endo, S. 1931. The host plants of *Hypochnus centrifugus* (Lev.) Tul. ever recorded in Japan. *Transaction Tottory Society of Agriculture Science* 3 : 254-270.

85. Fajola, A.O. 1978. The effects of some environmental factors on the reproductive structures of some species of *Cercospora. Nova Hedwigia* 29 : 922-934.

86. Franje, N.S. 1972. The effect of six fungicides on anthracnose of *Phaseolus aureus*. B.S. Thesis, College of Agriculture, UPLB, Laguna Philippines.

87. Gangadharan, K., Jeyarajan R. and Rabindran R. 1989. Diseases of pulses and their management in Tamil Nadu. *National Symposium on New Frontiers in Pulses Research and Development, Kanpur* 120-121 pp.

8. Ghani Arif, A. and Munjal R.L. 1955. The causal fungus of the blight disease of *Phaseolus* in the Punjab. *Pakistan Journal of Science & Research* 7 (3) : 83-92.

9. Ghosh, T., Mukherji N. and Basak M. 1964. On the occurrence of new species of *Orbilia. Jute Bulletin* 27 : 134-141.

0. Gill, K.S., Sandhu T.S., Singh K. and Brar J.S. 1975. Evaluation of mungbean (*Vigna radiata* (L.) Wilczek) germ plasm. *Crop Improvement* 2 : 99-104.

1. Golato, C. 1966. Bean rust in Somalia. *Rivista de Agricoltura Subtropicale e Tropicale* 60 (4/6) : 178-181.

2. Golato, C. 1967. Cercosporiosi del Fagiolo in Somalia. *Rivista de Agricoltura Subtropicale e Tropicale* 61 : 159-162.

3. Grewal, J.S. 1977. Diseases of Mungbean in India. *First International Mungbean Symposium, AVRDC, Taiwan*, 165-168 pp.

4. Grewal, J.S. 1988. Diseases of Pulse crops - an overview. *Indian Phytopathology* 41 : 1-14.

5. Grewal, J.S., Pal M. and Kulshrestha D.D. 1980. Control of Cercospora leaf spot of greengram by spraying bavistin. *Indian Journal of Agricultural Sciences* 50 : 707-711.

6. Grewal, J.S., Pal M. and Kulshrestha D.D. 1988. Fungal diseases of Pulse crops. In "Pulse Crops" (B. Baldev, S. Ramanujam and H.K. Jain Eds.) Oxford and IBH Publishing Co. Pvt. Ltd. New Delhi, 259-315 pp.

7. Grover, R.K. and Chopra B.L. 1977. Seed soil and foliage treatments with carboxin and oxicarboxin for control of *Rhizoctonia* sp. and their toxicity to other fungi. *Pesticides* 11 (9) : 38-41.

8. Grover, R.K. and Sakhuja P.K. 1981. Some pathological studies on *Rhizoctonia bataticola* leaf blight of mungbean. *Indian Phytopathology* 34 : 24-29.

9. Gupta, B.M. 1970. New disease on mungbean from Udaipur Rajasthan, India. *Plant Disease Reporter* 54 : 453.

00. Gupta, B.M. and Reddy G.S. 1975. Physiological studies on *Alternaria alternata* (Fr.) Keissl. the incitant of leaf spot of mungbean (*Phaseolus aureus* Roxb.). *Labdev B* 13 : 217-218.

01. Gupta, R.B.L., Singh G., Singh R.R. and Solanki J.S. 1975. Efficacy of different fungicides against powdery mildew of mung. *Indian Phytopathology* 28 : 164-166.

102. Gupta, S.C. and Sinha S. 1951. Further additons to the *Synchytria* o India. *Indian Phytopathology* 4 : 7-10.

103. Gupta, V.K. and Saharan G.S. 1974. Epidemiological observation o leaf spot (*Cercospora* spp.) of blackgram. *Indian Phytopathology* 27 606-608.

104. Gurha, S.N. 1981. Screening blackgram germplasm types for lea blight disease (*Macrophomina phaseolina*) resistance. *Madra Agricultural Journal* 68 : 692.

105. Gurha, S.N. and Gangal L.K. 1980. Control of powdery mildew o greengram (*Vigna radiata* (L.) Wilczek). *Madras Agricultural Journal* 67 666-668.

106. Gurha, S.N. and Misra D.P. 1983. A note on the field reaction o Cercospora leaf spot in blackgram (*Vigna mungo*). *Indian Journal c Mycology and Plant Pathology* 13 : 351-352.

107. Hamdollah-Zadeh, A. 1987. Powdery mildew of mungbean in Irar *Iranian Journal of Plant Pathology* 23 (1-4) : 37, 103.

108. Hans, J.K., Tyagi P.D., Kataria H.R. and Grover R.K. 1981. Th influence of soil and other physical factors on the antifungal activity c carbendazim against *Rhizoctonia solani*. *Pesticides* 12 (4) : 425-432.

109. Hansford, C.G. 1943. Contributions towards the fungal flora c Uganda. V. Fungi Imperfecti. *Proceedings of Linean Society Londor* 34-67 pp.

110. Haware, M.P. 1972. A note on the angular black spot of *Phaseolu mungo* and *Phaseolus radiatus*. *JNKVV Research Journal* 6 : 55-56.

111. Haware, M.P. and Pavgi M.S. 1969. Laboratory evaluation of som fungicides and antibiotics for the control of angular black spot c legumes. *Hidustan Antibiotic Bulletin* 12 (1) : 17-21.

112. Haware, M.P. and Pavgi M.S. 1969. Field trials with some fungicid and antibiotic aureofungin in control of angular black spot of blackgra and greengram. *Hindustan Antibiotic Bulletin* 12(1) : 22-25.

113. Haware, M.P. and Pavgi M.S. 1971. Perpetuation of *Protomycops* species causing purple leaf spot of legumes in India. *Annals of th Phytopathological Society of Japan* 37 : 242-248.

114. Haware, M.P. and Pavgi M.S. 1972. Physiology of chlamydospo germination in *Protomycopsis* species. *Friesia* 10 : 43-51.

115. Haware, M.P. and Pavgi M.S. 1976. Chlamydospore germination in *Protomycopsis* speices. *Mycopathologia* 59 : 105-111.

116. Haware, M.P. and Pavgi M.S. 1976. Cytology of chlamydospore development in *Protomycopsis* species. *Cytologia* 41 : 459-465.

117. Haware, M.P. and Pavgi M.S. 1976. Field reaction of blackgram and greengram to angular black spot. *Indian Journal of Agricultural Sciences* 46 : 280-282.

118. Haware, M.P. and Pavgi M.S. 1977. Host range of two *Protomycopsis* species from India. *Sydowia* 29 : 272-274.

119. Haware, M.P. and Pavgi M.S. 1977. Cytology of chlamydospore germination and *in vitro* development of *Protomycopsis* species. *Caryologia* 30 : 313-331.

120. Hiremath, R.V. and Shambulingappa K.G. 1981. Macrophomina stem blight of blackgram and its effect in some varieties. *Current Research* 10(1) : 11-13.

121. Hoffmaster, D.E., McLaughlin J.H., Ray W.W. and Chester K.S. 1943. The problem of dry rot caused by *Macrophomina phaseoli* (=*Sclerotium bataticola*). *Phytopathology* 33 : 1113-1114.

122. Hooda, I. and Grover R.K. 1982. Studies on different isolates, age and quantity of inoculum of *Rhizoctonia bataticola* in relation to disease development in mungbean. *Indian Phytopathology* 35 : 619-623.

123. Hooda, I. and Grover R.K. 1983. Nutritionally mediated virulence of *Rhizoctonia bataticola* on mungbean (*Vigna radiata*). *Indian Phytopathology* 36 : 662-666.

124. Hooda, I. and Grover R.K. 1983. Comparative anti-fungal activity of fungitoxicants against *Rhizoctonia bataticola* causing seedling rot and foliage blight of mungbean. *Indian Journal of Plant Pathology* 1 : 75-82.

125. Hooda, I. and Grover R.K. 1983. Fungicidal control of *Macrophomina phaseolina* altered in pathogenicity by substrate nutrients. *Annals of Applied Biology* 104 : 69-78.

126. Hooda, I. and Grover R.K. 1988. Effect of age, quantity of inoculum and isolates of *Macrophomina phaseolina* on the pathogenesis of mungbean and its control by chemicals. *Indian Phytopathology* 41 : 107-117.

127. Hossain, I., Shaikh M.A.Q. and Khan A.A. 1981. Field screening of mungbean (*Vigna radiata* (L.) Wilczek) germplasm for resistance to Cercospora leaf spot in Bangladesh. *SABRAO Journal* 13 : 180-182.

128. Houston, B.R. 1945. Culture types and pathogenicity of isolates of *Corticium solani*. *Phytopathology* 35 : 371-393.

129. Ilag, L.L. 1977. Fungal diseases of mungbean in the Philippines. *First International Mungbean Symposium, AVRDC, Taiwan,* 154-160 pp.

130. Ilag, L.L. 1977. Mungbean rust in the Philippines: Symptomatology and etiology. *kalikasan* 6(3) : 199-202.

131. Ilag, L.L. and Marfil V.E. 1977. Diplodia pod rot of mungbean. *Philippines Agriculturist* 61 : 186-191.

132. Jaganathan, T., Narayanaswamy P., Palanisamy A. and Ranganathan K. 1974. Studies on the damping off disease of blackgram (*Phaseolus mungo) (Pythium aphanidermatum)*. *Madras Agricultural Journal* 61 : 156-159.

133. Jain, N.K. 1970. Studies on *Rhizoctonia bataticola* (Taub.) Butler causing diseases of urid (*Phaseolus mungo* L.). M.Sc. Ag. Thesis, JNKVV, Jabalpur, M.P., India, 60 pp.

134. Jain. N.K. and Khare M.N. 1972. Chemical control of *Rhizoctonia bataticola* causing diseases of urid. *Mysore Journal of Agricultural Sciences* 6 : 461-465.

135. Jain, N.K., Khare M.N. and Sharma H.C. 1973. Variation among the isolates of *Rhizoctonia bataticola* from urid plant parts and soil. I. In pathogenicity, morphology and growth pattern. *Mysore Journal of Agricultural Sciences* 7 : 411- 413.

136. Jain, N.K., khare M.N. and sharma H.C. 1972. Variation among the isolates of *Rhizoctonia bataticola* from urid plant parts and soil. II. Temperature and pH requirements. *JNKVV Research Journal* 6: 165-166.

137. Jalali, B.L., Khirbat S.K. and Sangwan M.S. 1981. Studies on different aspects of disease resistance in pulse crops. *Third International Symposium on Plant Pathology,* IPS, *New Delhi*, 161 p.

38. Jalali, Indu and Grover R.K. 1979. Uptake of chloroneb by urd bean and its use as seed treatment for disease control. *Indian Phytopathology* 32 : 564-567.

139. Jhooty, J.S. and Bains S.S. 1972. Evaluation of different systemic and non systemic fungicides for the control of damping off of mung *(Phaseolus aureus)* caused by *Rhizoctonia solani. Indian Phytopathology* 25 : 509-512.

140. Jhooty, J.S. and Bains S.S. 1976. Studies on infection cushion formation in *Rhizoctonia solani. Indian Phytopathology* 29 : 303-304.

141. Jhooty, J.S. and Singh H. 1972. Stability of benomyl in plants. *Phytochemistry,* 11 : 2207-2208.

142. Joshi, R.K. 1964. Comparative pathogenicity of two isolates of *Rhizoctonia bataticola on Phaseolus aureus.* M.Sc. Ag. Thesis, JNKVV, Jabalpur, M.P., India, 42 pp.

143. Kaiser, W.J. 1970. *Rhizoctonia* stem canker disease of mungbean (*Phaseolus aureus*) in Iran. *Plant disease Reporter* 54 : 246-250.

144. Kaiser, W.J., Danesh D., Okhovat M. and Mossahebi H. 1968. Regional Pulse Improvement Project : Diseases of pulse crops (edible legumes) occurring in Iran. *Iranian Journal of Plant Pathology* 4(3):2-6.

145. Kaiser, W.J., Okhovat M., Mossahebi G.H., Danesh D., Zonuzi A. and Mojahad P. 1969. Plant Pathology. In progress report of regional pulse improvement project, U.S. Department of Agriculture, ARS, US Agency for International Development No. 7, 47-80 pp.

146. Kaiser, W.J., Mossahebi G.H. and Okhovat M. 1970. Occurrence, pathogenicity and distribution in soil of *Rhizoctonia solani* inciting a stem canker disease of mungbean (*Phaseolus aureus*) in Iran. *Iranian Journal of Plant Pathology* 6 : 17-25.

147. Kajiwara, R. and Mukelar A. 1975. Occurrence of mungbean scab in Indonesia. *Proceedings Third Congress of Indonesian Phytopathological Society, Bogor, Indonesia,* 22 to 25 Feb.

148. Kajiwara, R. and Mukelar A. 1976. Mungbean scab caused by *Elsinoe* in Indonesia. *Contrib. Centr. Research Institute Agriculture, Bogor No.* 23 : 1-12.

149. Kajiwara, T. and Mukelar A. 1976. Mungbean scab caused by *Elsinoe* in Indonesia. *Biological abstract 66* :37862.

150. Kamthan, K.P.,Gurha S.N. and Misra, D.P. 1981. Fungicidal control of Cercospora leaf spot of urd bean *(Vigna mungo* (L.) Hepper). *Madras Agricultural Journal* 68 : 604- 605.

151. Kannaiyan, S., Venkatarao A. and Thangamani G. 1974. Control of *Cercospora* leaf spot of greengram. *Labdev* B 12 : 150-151.

152. Kartha, K.K. 1965. Rhizoctonia disease of mung, Factors affecting the incidence and severity of the disease. M.Sc. Ag. Thesis, JNKVV. Jabalpur, M.P., India.

153. Kartha, K.K. and Nema K.G. 1969. Effect of the host nutrition on the incidence and Severity of *Rhizoctonia* disease of *Phaseolus aureus. Indian Phytopathology* 22 : 471-75.

154. Kataria, H.R. and Grover R.K. 1975. Inhibition of infection structure of *Rhizoctonia solani* by fungitoxicants. *Zeitschrift Pflanzenkrankheiten und Pflenzenschutz* 82 : 226-232.

155. Kataria, H.R. and Grover R.K. 1976. Fungitoxicity of mineral oils against *Rhizoctonia solani* causing damping off of mungbean (*Phaseolus aureus*). *Annals of Applied Biology* 83 : 79-85.

156. Kataria, H.R. and Grover R.K. 1977. Comparison of fungicides for the control of *Rhizoctonia solani* causing damping off of mungbean (*Phaseolus aureus*). *Annals of Applied Biology* 88 : 257- 265.

157. Kaushal, R.P. and Singh B.M. 1988. Genetics of disease resistance in urdbean (*Vigna mungo* (L.)Hepper) to the leaf spot caused by *Colletotrichum truncatum* (Schw) Andrus & Moore. *Euphytica* 37 : 279-281.

158. Kaushik, C.D., Chand J.N. and Satya Vir. 1981. Investigations on seed borne nature of *Rhizoctonia bataticola* causing leaf blight of mungbean. *Third International Symposium on Plant Pathology, IPS, New Delhi,* 87 pp.

159. Kernkamp, M.F. and Hemeric G.A. 1953. The relation of *Ascochyta imperfecta* to alfalfa seed production in Minnesota. *Phytopathology* 43: 378-383.

160. Khandar, R.R., Bhatnagar M.K. and Rawal P.P. 1983. Evaluation of some commercial varieties of mungbean (*Vigna radiata*) against the leaf spot caused by *Cercospora canescens. Agricultural Science Digest India* 3: 182-184.

161. Khandar, R.R., Bhatnagar M.K. and Rawal P.P. 1985. Cultural conditions affecting growth and sporulation of *Cercospora canescens,* incitant of leaf spot of mungbean and germination of its spores. *Indian Journal of Mycology and Plant Pathology* 15 : 165-171.

162. Khandar, R.R., Bhatnagar M.K. and Rawal P.P. 1986. Chemical control of mungbean leaf spot incitant (*Cercospora canescens* Ellies & Martin) under laboratory and green house conditions. *Pesticides* : 20 (3) : 45-48.

163. Khare, M.N., Jain N.K. and Sharma H.C. 1970. Variation among the *Rhizoctonia bataticola* isolates from urid bean plant parts and soil. *Phytopathology* 60 : 1298.

164. khare, M.N. and Shree Kumar K. 1980. Role of rain splash in the transmission of seed-borne pathogens to phyllosphere of urd bean plants. *Indian Phytopathology* 33 : 152-153.

165. Khare, M.N. and Singh V. 1981. Three seed borne *Colletotrichum* spp in uridbean, their detection and pathogenic potential. *Third International Symposium on Plant Pathology, IPS, New Delhi,* 87 pp.

166. Khosala, H.K., Naik S.L., Mandloi S.C. and Goray S.C. 1988. Control of powdery mildew of mung and urid in relation to losses and disease development. *Indian Phytopathology* 41 : 59-63.

167. Kishore, N., Asthana A. and Dubey N.K. 1987. Antifungal activity of rhizome vapours of *Alpina carinata* against *Rhizoctonia solani*. *Transaction of British Mycological Society* 88 : 136-138.

168. Kobayashi, K. and Ui T. 1977. Wilt inducing antibiotic compounds produced by *Cephalosporium gregatum*. *Physiological Plant Pathology* 11 : 55-60.

169. Kolte, S.J. and Shinde P.A. 1973. Influence of plant extract of certain hosts on the growth and sclerotial formation of *Macrophomina phaseoli in vitro*. *Indian Phytopathology* 26 : 351-352.

170. Konde, B.K., Shinde D.B. and more B.B. 1980. Studies on root infecting fungi and *Rhizobium* of blackgram (*Vigna mungo* (L.) Hepper). *Journal of Maharashtra Agricultural Universities* 5 : 222-225.

171. Konno, Shoshin and Narikawa T. 1977. Recent studies and problems on breeding and cultivation of adzuki bean in Japan. *First International Mungbean Symposium, AVRDC, Taiwan* 236-239 pp.

172. Kotasthane, S.R. and Agrawal S.C. 1976. control of foliar diseases of mungbean (*Phaseolus aureus*) by fungicides. *Pesticides* 10(8) : 35-36.

173. Kotasthane. S.R. and Gupta Om. 1985. Economics of systemic and non systemic fungicides in control of diseases of moong. *JNKVV Research Journal* 17 : 158-160.

174. Koul. A.K. 1980. Two new hosts for *Dactuliophora. Indian Phytopathology* 33 : 622.

175. Kovalenko, E.D. 1967. Intergeneric specialization of *Ascochyta* from soybean. *Zashch. Rast. Mosk* 12(8) : 46.

176. Kraft, J.M. 1967. Effect of nutrients and inoculum density on the virulence of *Pythium aphanidermatum and Pythium ultimum* to mungbean seedlings. *Disease Abstract* 27 (11) B : 3744-3745.

177. Kraft, J.M. and Erwin D.C. 1967. Stimulation of *Pythium aphanidermatum* by exudates from mungbean seeds. *Phytopathology* 57: 866-868.

178. Kraft, J.M. and Erwin D.C. 1968. Effect of inoculum substrate and density on the virulence of *Pythium aphanidermatum* to mungbean seedlings. *Phytopathology* 58 : 1427-28.

179. Krishnaiah, V.V., Venkateswarlu P.. Prasad M.M.K.D. and Rao Y.H. 1978. A new strain of blackgram (*Vigna mungo* (L.) Hepper) resistant to powdery mildew (*Erysiphe polygoni). Tropical Grain Legume Bulletin* 13/14 : 28-29.

180. Krishnamohan, G., Gangadharan K. and Shanmugasundaram P. 1986. Efficacy of certain fungicides against powdery mildew disease of blackgram (*Vigna mungo* L.). *Madras Agricultural Journal* 73 : 173-175.

181. Kumar, S.M., Khare M.N. and Srivastav S.K. 1969. *Macrophomina* leaf spot of urid (*Phaseolus mungo*). I. Some observations on disease resistance. *Mysore Journal of Agriculture Sciences 3* : 472 - 474.

182. Kush, A.K. 1982. Interaction between symbiosis and root pathogenesis in greengram (*Vigna radiata* (L.) Wilczek). *Plant and Soil* 65 : 133-135.

183. Kwon, S.H. and OH J.H. 1981. Sporulation of *Cercospora canescens* Ellis & Martin in culture. *Korean Journal of Plant Protection* 20 : 21 -24.

184. Lai, M.T., Weinhold R. and Handcock J.G. 1968. Permeability changes in *Phaseolus aureus* associated with infection by *Rhizoctonia solani. Phytopathology* 58 : 240-245.

185. Leach, R. 1932. Report of the Mycologist. Annual Report department of Agriculture, Nyasaland, 47-50 pp.

186. Legaspi, B.M., Catipon E.M. and Hubbell J.N. 1977. AVRDC philippine outreach Program, Mungbean studies. *First International Mungbean Symposium, AVRDC, Taiwan* 220-223 pp.

187. Lilhore, D.R. 1988. Epidemiological studies on leaf blight of urd bean caused by *Rhizoctonia bataticola*. M.Sc.Ag. Thesis, R.A.K. College of Agriculture, Sehore, M.P., India, 87 p.

188. Liu, H.Y. 1961. Citrus gummosis due to *Rhizoctonia sp*. *Special Publication College of Agriculture Taiwan University* 10 : 161-177.

189. Litzenberger, S.C. and Stevenson J.A. 1957. A preliminary list of Nicaraguan plant diseases. *Plant Disease Reporter* Supplement 243 : 19 pp.

190. Mahmood, M., Kumar S. and Prasad H. 1976. Bavistin, a systemic fungicide as foliar spray for the control of Cercospora leaf spot of mung and urd. *Proceedings Bihar Academy of Agricultural Sciences* 24(2) : 101-102.

191. Majid, S. 1953. *Annual Reports of the Grow More Food Activities of the Department of Agriculture Assam* 1951-52 and 1952-53. Part II, 107 pp.

192 Malhan, I., Tyagi P.D. and Grover R.K. 1973. Systemic activity of benomyl in urd bean(*Phaseolus mungo* L.). *Acta Phytopathologica Academiae Scientiarum Hungaricae* 8 : 295-300.

193. Malhan, I., Tyagi P.D. and Grover R.K. 1975. Physical factors affecting the toxicity of benomyl to *Rhizoctonia solani in vitro*. *Indian Phytopathology* 28 : 491-494.

194. Mallaiah, K.V., Vijayalakshmi M. and Rao A.S. 1981. New records of some foliar diseases. *Indian Phytopathology* 34 : 247.

195. Mathur, A.K. and Tyagi R.N.S. 1983. A new pod rot due to *Pythium debaryanum* in cowpea, mung and urd. *Indian Journal of Mycology and Plant Pathology* 13 : 219 - 220.

196. Mathur, A.K. and Tyagi R.N.S. 1984. Occurrence of *Choanephora* pod rot on Kharif pulses in Rajasthan. *Indian Journal of Mycology and Plant Pathology* 14: 152.

197. Mathur, R.L., Daftari L.N. and Jhamaria S.K. 1971. Effectiveness and economics of fungicidal control of powdery mildew of urid (*Phaseolus mungo L.*). *Rajasthan Journal of Agriculture Science* 2: 21-23.

198. Mayee, C.D. and Datar V.V. 1986. "Phytopathometry". *Technical Bulletin -1, Marathawada Agricultural University, Parbhani, India* 146. pp.

199. Mathur, R.S. 1954. Diseases of Pulse crops in Uttar Pradesh, *India. Agriculture & Animal Husbandry* 5: 24-28.

200. McRae, W. 1929. New diseases reported during the year 1928 from India. *International Bulletin of Plant Protection* 3 (2) : 21-22.

201. McRae, W. 1933. Report of Imperial Mycologist. *Scientific Report, Imperial Institute of Agriculture Research Pusa, 1931-32,* 122-140 pp.

202. Menancio, D.I. and Ramirez D.A. 1977. Genetic polymorphism and ontogenetic isozyme patterns of Cercospora leaf spot resistant and susceptible varieties of two *Vigna* species. *First International Mungbean Symposium, AVRDC, Taiwan,* 149-153 pp.

203. Menancio, D.I. and Ramirez, D.A. 1977. Genetic polymorphism and ontogenetic isozyme patterns of *Cercospora* leaf spot resistant and susceptible varieties of mungbean. *Philippine Journal of Crop Sciences* 2: 197-202.

204. Messiaen, C.M. and Beyries A. 1965. Estimation of the infestaion of soil samples by virulent strains of *Corticium solani*. Application to the study of the effect of organic amendments. *Phytiat. Phytopharm.* 14 : 183-191.

205. Mew. I.P., Wang T.C. and Mew T.W. 1975. Inoculum production and evaluation of mungbean varieties for resistance to *Cercospora canescens. Plant Disease Reporter* 59: 397-611.

206. Mishra, R.P., Khare M.N. and Chand J.N. 1971. *Sclerotium* leaf spot of urid (*Phaseolus mungo L.*). *Indian Phytopathology* 24: 609-611.

207. Moghe, S.V. and Utikar P.G. 1981. Effect of sowing dates on the incidence of powdery mildew disease on greengram (*Vigna radiata). Indian Journal of Mycology and Plant Patholoogy* 11 : 98-99.

208. Moghe, S.V., Utikar P.G. and More B.B. 1982. Fungicidal control of powdery mildew (*Erysiphe polygoni* DC) of greengram (*Vigna radiata* (L.) Wilczek). *Pesticides* 16 (8): 10-11.

209. Mohan, R., Subramanian C.L. and Marimuthu T. 1980. Control of rust disease of greengram. *MACCO Agricultural Digest Research Bulletin* 5 (6) : 14.

210. Mukelar, A. 1977. Mung bean scab in Indonesia. *First International Mungbean Symposium, AVRDC, Taiwan,* 161-164. pp.

211. Mukelar, A., Sudjadi M. and Kajiwar T. 1976. Chemical control for mungbean scab. *Contributions Central Research Institute for Agriculture* No. 24 : 7 pp.

212. Muller, A.S. and Chupp C. 1945. The Cercosporae of Venezuela. *Bol. Society Yenezol, Cien. nat.* 8: 35-49

213. Muller, A.S. 1953. A foliar disease of legumes in Central America. *FAO Plant Protection Bulletin* 6: 83-84.

214. Munjal, R.L. 1960. A commonly occurring leaf spot disease caused by *Myrothecium roridum*. *Indian Phytopathology* 13: 150-155.

215. Muthusamy, S. and Ragupathy N. 1983. Efficacy of fungicides in the control of blackgram rust. *Madras Agricultural Journal* 70: 345-346.

216. Muthusamy, S. and Ragupathy N. 1986. Studies on efficacy of vitavax in the control of blackgram rust. *Pesticides* 20 (1) : 29-30.

217. Nacien, C.C. 1924. Studies on *Rhizoctonia* blight of beans. *Philippine Agriculturist* 8 : 315-321.

218. Naik, S.L. Khosla H.K. and Mandloi S.C. 1984. Control of powdery mildew of mung and urid in relation to losses and disease development. *Indian Phytopathology* 37 : 383.

219. Nalampang, A. 1977. Mungbean production in Thailand. *First International Mungbean Symposium, AVRDC, Taiwan* 12-14 pp.

220. Narasimhan, V., Cherulu. C. and Satyanarayana A. 1989. Sources of resistance against angular black leaf spot disease in mungbean. *Indian Journal of Pulses Research* 2: 184.

221. Nattrass, R.M. 1941. Plant diseases in Kenya during 1940. *East African Agriculture Journal* 7(1) : 57.

222. Nawaz, R.M.S. and Narayanaswamy P. 1983. Chemical control of powdery mildew disease of blackgram and greengram. *Pesticides* 17 (2): 23-24, 26.

223. Nawaz, R.M.S. and Narayanaswamy P. 1983. Effect of powdery mildew infection on growth and yield of blackgram. *Madras Agricultural Journal* 70 : 179-181.

224. Nawaz, R.M.S. and Narayanaswamy P. 1983. Influence of host nutrition on powdery mildew disease development in blackgram. *Madras Agricultural Journal* 70 : 57-58.

225. Nawaz R.M.S. and Narayanaswamy P. 1983. Effect of powdery mildew infection on the nitrogen metabolism of blackgram. *Madras Agricultural Journal* 70 : 316-318.

226. Nawaz, R.M.S. and Narayanaswamy P. 1983. Interaction on powdery mildew and leaf crinkle virus of blackgram. *Madras Agricultural Journal* 70: 487-488.

227. Niganova, M.S. 1962. The biology of the fungus *Ascochyta imperfecta,. Uzbek. Biol. Zh.* 6 (5) : 39-44.

228. Paik, S.B. and Do E.S. 1987. *Fusarium* spp. isolated from seed, root and cultivated soil of *Phaseolus* vi (ri) *dissimus* and their pathogenicity. *Korean Plant Pathology* 3 : 8-12.

229. Pande, A. 1985. Biocontrol characteristics of some moulds. *Biovigyanam* 11 (1) : 14-18.

230. Pandher, R.S. and Singh G. 1981. Studies on physiology, pathogenicity, symptomatology and morphology of *Cercospora canescens* and *Cercospora cruenta* of mungbean. *Third International Symposium on Plant Pathology, IPS, New Delhi,* 162-163. pp.

231. PANS. 1981. "Pest Control in Tropical Grain Legumes". *PANS Mannual No.* 5 : 206 pp.

232. Park, M. 1934. Report on the work of the Mycological Division, Ceylon Administration Reports. *Report of the Director of Agriculture 1933,* 126-133 pp.

233. Parmeter J.R., Sherwood R.T. and Platt W.D. 1969. Anastomosis grouping amongst isolates of *Thanatephorus cucumeris. Phytopathology* 59 : 1270-1278.

234. Patel. M.K., Kamat M.N. and Bhide V.P. 1949. Fungi of Bombay, Supplement I. *Indian Pytopathology* 2 : 142-155.

235. Patel, M.K., Kulkarni Y.S. and Dhande G.W. 1949. A new *Synchytrium* on *Phaseolus mungo. Current Science* 18: 171.

236. Patil, S.D. and Mahabale T.S. 1964. The genus *Synchytrium* in Maharashtra. *Journal of University of Poona* 26 : 67-79.

237. Patil, N.K., Adiver S.S., Hiremath P.C. and Hegde R.K. 1987. Assessment of blackgram varieties for resistance to powdery mildew and yellow mosaic virus. *Current Research* 16 (3) : 41-42.

238. Paul, Y.L. and Kaushal R.P. 1985. Effect of relative humidity on germination of powdery mildew conidia. *Indian Phytopathology* 38 : 757-758.

239. Paul, Y.S. and Kapoor J.N. 1983. A revision of *Erysiphe polygoni* sensu Salmon from India. *Indian Phytopathology* 36: 247-250.

240. Pavgi, M.S. and Haware M.P. 1970. Culture of two Protomycopsis species from India. *Nova Hedwigia* 19:323-328.

241. Pavgi, M.S. and Haware M.P. 1970. Conidial discharge in *Protomycopsis thirumalacharii. Pathol. et microbiol.* 35: 297-301.

242. Pavgi, M.S, and Thirumalachar M.J. 1953. Angular black spot disease of mung. *Nature London* 172: 314-315.

243. Pawar, V.H. and Thirumalachar M.J. 1970. Mechanism of parasitism in *Myrothecium roridum* Tode. *In "Plant Diseases Problems", Indian Phytopathological Society, IARI, New Delhi*, 173-178 pp.

244. Payak, M.M. 1951. A *Synchytrium* disease of urid bean. *Current Science* 20: 103-104.

245. Pedgaonkar, S.M. and Bhusari K.M. 1989. Comparative studies on foliar and root isolates of *Macrophomina phaseolina* causing leaf blight of mung and charcoal rot of Jowar. *Indian Phytopathology* 42 : 349.

246. Persons, T.D. 1934. New or unusual diseases reported or observed in Mississippi in recent years. *Phytopathology* 24 : 843.

247. Philip, C.T. 1963. Studies of *Rhizoctonia* disease of mung at Jabalpur. M.Sc. Ag. Thesis, JNKVV, Jabalpur., M.P., India, 70 pp.

248. Philip, C.T., Kartha K.K., Joshi R.K. and Nema K.G. 1969. A *Rhizoctonia* disease of mung (*Phaseolus aureus* Roxb.) in Madhya Pradesh. *JNKVV Research Journal* 3: 40-43.

249. Poehlman, J.M 1977. What we have learned from the international mungbean nurseries. *First International Mungbean Symposium. AVRDC, Taiwan,* 97-100 pp.

250. Poehlman, J.M., Sechler D.T., Yoke J.M., Watt E.E., Sevindell R.E. and Bashandi M.M.H. 1973. Special Report of University M.O. Agricultural Experiment Station, 158 pp.

251. Prasad, K.S., Rao P.R. and Wadia K.D.R. 1983. *Protomycopsis* leaf spot of greengram. *Indian Phytopathology* 36: 178.

252. Prasad, N., Agrawal, J.P. and Agnihotri J.P. 1962. The genus *Protomycosis* in India. *Indian Phytopathology* 15: 24-27.

253. Preston, D.A. 1946. Legume diseases previously unreported for Oklahoma. *Plant Disease Reporter* 30 : 45-46.

254. Preston, D.A., and Ray W.W. 1943. Yeast spot of soybeans and mungbeans caused by *Nematospora coryli* Peglion in Oklahoma. *Plant Disease Reporter* 27 : 601-602.

255. Purkayastha, R.P. 1975. Influence of colchicine on the disease susceptibility of gram seedlings to *Fusarium orthoceras*. *Science & Culture* 41: 546-547.

256. Quebral, F.C. 1969. Green house test for controlling mungbean powdery mildew by incorporating chemical in the soil. *Philippine Phytopathology* 5 : 52-54.

257. Quebral, F.C. 1977. Powdery mildew and *Cercospora* leaf spot of mungbean in the Philippines. *First International Mungbean Symposium, AVRDC, Taiwan,* 147-148 pp.

258. Quebral, F.C. and Cagampang I. 1970. Influence of *Cercospora* leaf spot control on yield of mungbean. *Agriculture at Los Banos* 9 (4) : 4-7, 12.

259. Quebral, F.C. and Lantican R.M. 1969. Effect of benlate on powdery mildew and yield of mungbean. *Agriculture at Los Banos*, 9 (1) : 13-14.

260. Quimio,T.H. 1975. Anthracnose organisms and disease of Philippines crops. *Technical Report No. 30. U.P. Natural Science Research Centre Diliman, Metro, Manila.*

261. Rachie, K. O. 1974. Grain legumes of the low land tropics: Mungbean In "Advances in Agronomy" (K.O. Rachie and L.M. Roberts) 26 : 62-77.

262. Raghunathan, V. 1968. Damping off of greengram, cauliflower dhaincha, ragi and clusterbean. *Indian Phytopathology* 21: 456-457.

263. Raghunathan, V. 1970. Root rot of *Arachis hypogea* seedlings caused by *Pythium debaryanum*. *Current Science* 39 : 44.

264. Rajak, R.C. and Rai M.K. 1982. Species of *Phoma* from legumes. *Indian Phytopathology* 35: 609-612.

265. Raj Kumar and Mukhopadhyay A.N. 1987. Field evaluation of urdbean germplasm lines against *Colletotrichum capsici*. *Indian Journal of Mycology and Plant Pathology* 17 : 66.

266. Ramakrishna Rao, S.V. and Chatrath M.S. 1989. Translocation of thiophanate methyl in mungbean (*Vigna radiata*). *Indian Phytopathology* 42: 294-295.

267. Ranganathan, K., Jaganathan T. Palaniswamy A. and Narayanaswamy P. 1973. *Rhizoctonia* root rot on blackgram, lablab and frenchbean *Madras Agricultural Journal* 60: 588.

268. Rangaswamy G. and Prasad N.N. 1959. A new seedling blight of *Phaseolus aureus* Roxb. and *Phaseolus mungo* L. *Indian Phytopathology* 12 : 184-185.

269. Rao, A.V., Ramdos N. and kannaiyan S. 1976. Effect of certain fungicides on the control of powdery mildew disease of blackgram. *Annamalai University Agriculture Research Journal* 6: 113-115.

270. Rao, P.G. and Raghunatha R.N. 1955. Preliminary observations on the control of powdery mildew in blackgram and greengram. *Andhra Agriculture Journal* 2(6): 311.

271. Rao, V.G. 1965. The genus *Phyllosticta* in India. *Sydowia* 19 : 117-120.

272. Rao, V.G. 1965. *Alternaria tenuis* Auct in Bombay, Maharashtra. *Mycopathologia et Mycologia Applicata* 27 : 157-163.

273. Rao, V.G. 1969. Two new leaf spot diseases of economic plants from India. *Sydowia* 23: 192-193.

274. Rashid, A.Q.M.B., Barma A.C. and Shaikh M.A.Q. 1983. Seed borne fungi of mungbean and their pathogenicity. *Bangladesh Journal of Botany* 12 : 223-224.

275. Rath, G.C. and Grewal J.S. 1973. A note on *Cercospora* leaf spot of *Phaseolus aureus*. *Indian Journal of Mycology and Plant Pathology* 3: 204-207.

276. Rath, G.C. and Mishra D. 1975. Varietal reaction of *Phaseolus aureus* to *Sclerotium rolfsii*. *Indian Journal of Mycology and Plant Pathology* 5: 216.

277. Rawala, G.S., Rewal H.S. and Chahal S.S. 1977. Organic growth factors and trace elements for *Cercospora dolichi* and their gross effects on growth. *Indian Phytopathology* 30: 189-194.

278. Reddy, D.B. (Editor). 1971. Outbreaks of pest and diseases and new records. *Quarterly Newsletter Plant Protection Committee for the South East Asia and Pacific Region F.A.O* 14 (3-4) : 5-15.

279. Reddy, G.S. and Gupta B.M. 1974. Epidemiological studies of leaf spo of mungbean (*Phaseolus aureus* Roxb.) caused by *Alternaria alternata* (Fr.) Keissl. *Labdev B* 12(3) : 117.

280. Reddy, K.S., Pawar, S.E. and Bhatia C.R. 1987. Screening fo powdery mildew (*Erysiphe polygoni* DC) resistance in mungbean (*Vigna radiata* (L). Wilczek) using excised *leaves.Proceedings Indian Academy of Sciences (Plant science)* 97 : 365-369.

281. Reeder, R.T., Whitmarsh J., Gray L.E. and Peterson R.E. 1986 Inhibition of photosynthetic electron transport by metabolites produced by *Phialophora gregata. Physiological and Molecular Plant Patholog* 28 : 371-379.

282. Reinking, O.A. 1918. Philippine economic plant diseases. *Philippin Journal of Sciences* 13 : 165-216.

283. Reinking O.A. 1919. Plant Diseases in the Philippines. *Phytopatholog* 9 : 114-140.

284. Rewal, H.S. and Bedi P.S. 1976. Epidemiology and control o *Cercospora* leaf spot of mung in Punjab. *Indian Phytopathology* 29 102-103.

285. Ribeiro Do Vale, F.X., Chaves G.M. and Zambolin L. 1985. Host rang study of soybean rust in Brazil. *Soybean Rust Newsletter* 7 : 7-9.

286. Roy, A.J. 1967. Some fungi from Almora. *Indian Phytopathology* 2(340-344.

287. Roy, A.K. 1973. Host range of *Sclerotinia sclerotiorum* and *Sclerotiu rolfsii* in Jorhat, Assam. *Science and Culture* 39 : 319-320.

288. Saikia, U.N. 1976. Blight of mung caused by *Corticium sasakii* a ne disease recorded from Assam. *Indian Phytopathology* 29 : 61-62.

289. Saikia, U.N. and Phookan A.K. 1983. Efficacy of different fungicide against *Corticium sasakii*, the causal organism of blight of mun *Indian Phytopathology* 36: 752-754.

290. Saikia, U.N. and Roy A.K. 1976. Natural occurrence of *Corticium sasakii* on few plants in Jorhat, Assam. *Science & Culture* 42 : 228-229.

291. Saksena, H.K. 1979. Epidemiology of diseases caused by *Rhizoctonia* spp. *In Proceedings of Consulatants Group Discussion on the Resistance to Soil Borne Diseases of Legumes ICRISAT, Hyderabad* 59-64 pp.

292. Saksena H.K. 1985. Relationship of environment and *Rhizoctonia* aerial blight. *Indian Phytopathology* 38 : 584.

293. Saksena, H.K. and Dwivedi R.P. 1973. Web blight of blackgram caused by *Thanatephorus cucumeris. Indian Journal of Farm Sciences* 1(1) : 58-61.

294. Samyappan, R. Jeyarajan R. and Vidhyasekaran P. 1989. Management of soil borne diseases of pulse crops with *Trichoderma* spp. *National Symposium on New Frontiers of Pulses Research and Development Kanpur* 112 pp.

295. Samiyappan, R. and Vidhyasekaran P. 1981. Differences between *Macrophomina phaseolina* isolates causing root and leaf blight of uridbean. *Indian Phytopathology* 34 : 407-409.

296. Satischandra, K.M., Hiremath R.V. and Hegde R.K. 1979. Effect of organic amendments and fungicides on the saprophytic activity of *Rhizoctonia batatiocola* causing root rot of beans. *Indian Phytopathology* 32 : 543-546.

297. Sattar, A. 1952. Connection of *Rhizoctonia bataticola* (Taub.) Butler the causal fungus of root rot of cotton and some other isolates of *Rhizoctonia bataticola* with the pycnidial stage of *Macrophomina phaseoli* (Maubl.) Ashby.*Pakistan Journal of Scientific and Industrial Research* 4 : 31-35.

298. Saxena, D.R., Sapru T, and Saxena R.R. 1989. Biochemical parameters associated with low incidence of *Macrophomina* blight in mungbean *National Symposium on New Frontiers of Pulses Research and Development, Kanpur*, 118-119 pp.

299. Saxena, R.M. 1984. Evaluation of infection percentage and crop loss estimates of some seed borne infections of greengram and blackgram in Uttar Pradesh. *Indian Journal of Plant Pathology* 2 : 146-148.

300. Saxena R.M. and Gupta J.S. 1981. Field survey for suspected seed transmitted diseases of mungbean and urdbean in Uttar Pradesh *Indian Phytopathology* 34 : 340-345.

301. Saxena R.M. and Sinha S. 1977. Abnormal isolates of seed borne *Colletotrichum truncatum* (Schw.) Andrus & Moore from India. *Current Science* 46 : 716

302. Saxena, R.M. and Sinha S. 1977. Seed borne infection of *Vigna mungo* in Uttar Pradesh. *Indian Phytopathology* 30 : 582-583.

303. Saxena R.M. and Sinha S. 1977. Seed borne infections of *Vigna radiata* (L.) Wilczek var. *radiata* in Uttar Pradesh new records. *Science and Culture* 44 : 377-379.

304. Saxena, R.M. and Sinha S. 1978. Histopathological investigations of the field fungi associated with the seeds of *Vigna radiata* and *Vigna mungo* (L.) from Uttar Pradesh and Haryana. *Indian Journal of Mycology and Plant Pathology* 8 : 61.

305. Sen Gupta, P.K. 1974. Diseases of major pulse crops in India. *PANS* 20 : 409-415.

306. Shahzad, S. and Ghaffar A. 1986. *Macrophomina phaseolina* on some new hosts. *FAO Plant Protection Bulletin* 34 (3) : 163.

307. Sharada, R, and Shetty H.S. 1987. Location and transmission of *Macrophomina phaseolina* in blackgram seeds. *Indian Phytopathology* 40 : 194-196.

308. Sharma, H.C., Khare M.N., Joshi L.K. and Kumar S.M. 1971. Efficacy of fungicides in the control of diseases of kharif pulses-mung and urid. *All India Workshop on Kharif Pulses*, 2 pp.

309. Sharma, O.P. and Tiwari Anamika 1975. Effect of chloroneb on *Pythium aphanidermatum. Indian Phytopathology* 28 : 115-117.

310. Sharma, O.P., Tiwari Anamika and Kulkarni S.N. 1975. Effect of seed treatment with systemic and non systemic fungicides on the control of seedling blight of mung *(Phaseolus aureus)* caused by *Rhizoctonia solani. Indian Phytopathology* 28 : 114-115.

311. Shirshikar, S.P., Shinde V.K., Deshmukh R.V., Ashtaputre J.U. and Zote K.K. 1989. Identification of source of resistance to *Macrophomina* blight in mungbean. *Indian Phytopathology* 42 : 349.

312. Shree Kumar, K. 1974. Studies on foliar fungal diseases of urid leading to their control. M.Sc. Ag. Thesis, JNKVV, Jabalpur, M.P., India, 84 pp.

313. Shukla, V.N. 1958. Sclerotial blight of mung and urd. *The Nagpur Agriculture College Magazine* 32 (1) : 30-32.

314. Sideris, C.P. 1929. Pythiaceous root parasites of various agricultural plants. *Phytopathology* 19 : 1140-1147.

315. Sideris, C.P. 1932. Taxonomic studies in the family Pythiaceae II. *Pythium. Mycologia* 24 : 14-61.

316. Singh, B.M., Saharan G.S., Sood A.K. and Shyam K.R. 1978. *Ascochyta* leaf spot of mash and cowpea. *Indian Phytopathology* 31 : 388-389.

317. Singh, B.M., Sood A.K. and Saharan G.S. 1978. Occurrence of a leaf spot and blight of blackgram caused by *Colletotrichum dematium. Indian Phytopathology* 31 : 100-101.

318. Singh, D.P. 1981. Breeding for resistance to diseases in greengram and blackgram. *Theoritical and Applied Genetics* 59 : 1-10.

319. Singh, D.V. and Mishra A.N. 1975. Varietal resistance of greengram (*Phaseolus aureus*) to *Cercospora* leaf spot. *Indian Journal of Mycology and Plant Pathology* 5 : 207.

320. Singh, D.V. and Singh R.R. 1976. Chemical control of *Cerococspora* leaf spot of green gram. *Indian Phytopathology* 29 : 337-339.

321. Singh, D.V. and Singh R.R. 1978. Field evaluation of fungicides for the control of *Cercospora* leaf spot of green gram. *Pesticides* 12 (6) : 28-29.

322. Singh, D.V., Singh R.R. and Mishra A.N. 1976. Search for sources of resistance to powdery mildew in blackgram. *Current Science* 45 : 541.

323. Singh, Gurdip, Kapoor Shashi and Singh K. 1987. Multiple disease resistance in mungbean with special emphasis on yellow mosaic virus. *Second International Symposium on Mungbean, Bangkok, Thailand* 25-26 pp.

324. Singh, K., Khatri H.L. and Bansal R.D. 1975. Chemical control of *Cercospora* leaf spot of *Phaseolus aureus. Indian Journal of Mycology and Plant Pathology* 5 : 108-109.

325. Singh, K., Virmani S.S. and Chohan J.S. 1973. Field screening of urid (*Phaseolus mungo*) varieties against maturity wilt phase of the dry root rot disease caused by *Macrophomina phaseoli* in Punjab. *Indian Journal of Mycology and Plant Pathology* 3 : 194-195.

326. Singh, R.A. and Gurha S.N. 1980. Disease problems of some important pulses and prospects of their control. *Pesticides Annual*, 81-85 pp.

327. Singh, R.A., Gurha S.N. and Sahambi H.S. 1981. Diseases of kharif pulses. *Indian Farming Special Issue on Pulses August 1981*.

328. Singh, R.P. and Singh D.R. 1973. A new leaf spot of urd *(Phaseolus mungo L.). Indian Journal of Farm Sciences* 1 : 115.

329. Singh, R.R. and Shukla P. 1986. Sources of resistance to leaf blight of blackgram caused by *Colletotrichum truncatum. Indian Journal of Plant Pathology* 4 : 105-107.

330. Singh, R.R. and Shukla P. 1986. Cultural studies on *Colletotrichum truncatum* causing anthracnose of blackgram. *Indian Journal of Mycology and Plant Pathology* 16 : 172-174.

331. Singh, R.R. and Shukla P. 1987. Amino acid changes in blackgram leaves infected with *Colletotrichum truncatum. Indian Phytopathology* 40 : 241.

332. Singh, R.S. 1982. "Plant Pathogens". Oxford & IBH Publishing Co. New Delhi, 443 pp.

333. Singh, R.S. and Sekhon P.S. 1988 . Effect of *Epicoccum purpurescens* on *Macrophomina phaseolina* causing seedling blight and charcoal rot of mung bean. *Indian Phytopathology* 41 : 248-249.

334. Singh R.S. and Singh S. 1985. Interaction of various isolates of *Epicoccum purpurescens* to *Macrophomina phaseolina* causing disease in mungbean. *Indian phytopathology* 38 : 599.

335. Singh, R.S. and Singh S. 1986. Interaction of some isolates of *Epicoccum purpurescens with Macrophomina phaseolina. Plant Disease Research* 1 : 35-40.

336. Singh R.S., Thind T.S. and Sekhon P.S. 1984. Biocontrol of charcoal rot of mungbean caused by *Macrophomina phaseolina. Indian Journal of Mycology and Plant Pathology* 14 : VII.

337. Singh, Surjeet and Srivastava M.P. 1985. Assessment of yield losses in mungbean due to *Myrothecium. Agricultural Science Digest India* 5 : 114-115.

338. Singh, S.A. 1984. Two fungi attacking pulse crops. *FAO Plant Protection Bulletin* 32 : 112.

339. Singh, S.D. and Naik S.M.P. 1976. Field control of powdery mildew of urid by fungicides. *Indian Journal of Mycology and Plant Pathology* 6 : 99.

340. Singh, S.N. and Agrawal S.C. 1987. Chemical control of *Myrothecium* leaf spot and pod blight of soybean. *Indian Journal of Plant Protection* 15 : 31-33.

341. Singh, S.N., Agrawal S.C. and Khare M.N. 1986. Efficacy and cost benifit ratio of certain fungicides against foliar diseases of mungbean *(Vigna radiata L.). Indian Journal of Plant Protection* 14 : 63-65.

342. Singh, S.N. and Srivastava S.K. 1987. Effect of phosphorus application on the severity of some important diseases of mungbean and urdbean. *Agriculture Science Digest* 7 : 47 - 51.

343. Singh, V. 1917. Studies on *Colletotrichum* Corda species associated with Urid *(Phaseolus mungo)* seed. M.Sc. Ag. Thesis, JNKVV, Jabalpur, M.P., India, 98 pp.

344. Sinha, B.K. 1989. Chemical control of *Cercospora* leaf spot of mungbean and urdbean. *National Symposium on New Frontiers of Pulses Research and Development, Kanpur,* 121-122 pp.

345. Sivaprakasam, K. 1983. Efficacy of some chemicals in the control of blackgram *(Vigna mungo* (L.) Hepper) powdery mildew *(Erysiphe polygoni* DC). *Madras Agricultural Journal* 70 : 95-96.

346. Sivaprakasam, K., Anabalagan K.R., Ramakrishna C. and Radhakrishnan T. 1983. Influence of N P K on *Cercospora* leaf spot incidence of mungbean. *Pulse Crops Newsletter* 3 : 52-53.

347. Sivaprakasam, K., Marimuthu T., Radhakrishnan T. and Vairavan K. 1981. Influence of date of sowing and spacing on the incidence of powdery mildew of greengram and black gram. *Madras Agricultural Journal* 68 : 65-67.

348. Sivaprakasam, K. and Marimuthu T. 1982. Fungicidal control of *Cercospora* leaf spot of mungbean. *Indian Phytopathology* 35 : 330-331.

349. Sivaprakasam, K., Pillayarsamy K., Rajamani A. and Rajgopalan C.K.S. 1976. Note on the evaluation of black gram *(Phaseolus mungo) and greengram (Phaseolus aureus)* varieties for resistance to powdery mildew *(Erysiphe polygoni). Madras Agricultural Journal* 63 : 139-140.

350. Sohi. H.S., Jain S.S. Shrama S.L. and Varma B.R. 1964. New records of plant diseases from Himachal Pradesh. *Indian Phytopathology* 17 : 42-45.

351. Sohi. H.S., Shrama S.L. and Nayar S.K. 1972. New records of fungi from Himachal Pradesh. IV. *Research Bulletin Punjab Vniversity* 23 : 109-111.

352. Sohi, H.S., Sharma S.L. and Sachdev K.B. 1966. Powdery mildew from Himachal Pradesh I. *Indian Phytopathology* 19 : 235-237.

353. Solhein, W.G. and Stevens F.L. 1931. *Cercospora* studies II. Some tropical *Cercosporae. Mycologia* 23 : 365-405.

354. Sommatmadja, S. and Sutaraman T. 1977. Present status of mungbean breeding in Indonesia. *First International Mungbean Symposium, AVRDC, Taiwan,* 230-232 pp.

355. Sood, A.K., Shyam K.R. and Singh B.M. 1981. Comparative epidemiological competence of *Cercospora canescens and Cercospora cruenta* in Urdbean. *Third International Symposium Plant Pathology, IPS, NEW Delhi,* 111 pp.

356. Soria, J.A. and Quebral F.C. 1973. Occurrence and development of powdery mildew on mongo. *Philippine Agriculturist* 57 : 153-177.

357. Sprague, R. 1935. *Ascochyta boltshauseri* on beans in Oregan. *Phytopathology* 25 : 416-420.

358. Srivastava, M.P. 1980. Occurrence of *Myrothecium* leaf spot of mungbean in Haryana. *Indian Phytopathology* 33 : 137.

359. Srivastava, M.P. and Singh Surjeet. 1985. Studies on survival of *Myrothecium roridum* Tode ex. Fr.. *National Academy of Sciences Letters, India* 8 (1) : 3-4.

360. Srivastava, R.C., Khan M.S. and Lal M.P. 1978. Some new leaf spot fungi from India. *Indian Phytopathology* 31 : 524-525.

361. Staples, R.R. 1958. Report of the Department of Research and Specialist Services Southern Rhodesia for the year 1957. *Report Ministry of Agriculture Rhodesia Nyasaland 1956-57,* 7-86 pp.

362. Subramanian, C.L., Marimuthu T. and Mohan R. 1980. Note on the fungicidal control of *Cercospora* leaf spot of green gram. *MACCO Agricultural Digest* 5 (8) : 11.

363. Subramanian, C.V. 1971. "Hyphomycetes". *Indian Council of Agriculture Research, New Delhi,* 930 pp.

364. Sud, V.K. and Singh B.M. 1984. Effect of sowing date and row spacing on the development of leaf spot *(Cercospora canescens)* in urdbean. *Indian Phytopathology* 37 : 288-293.

365. Sud, V.K. and Singh B.M. 1984. Effect of environmental factors on the development of the leaf spot *(Cercospora canescens)* in urdbean. *Indian Phytopathology* 37 : 511-515.

366. Sud, V.K. and Singh B.M. 1985. Relationship between the severity of *Cercospora canescens* on urdbean *(Vigna mungo)* and loss in grain yield. *Indian Phytopathology* 38 : 435-441.

367. Sundaraman, S. 1931. Administration Report of the Mycologist, Madras for the year 1929-30, 30 pp.

368. Sundaraman, S. 1932. Administration Report of the Mycologist, Madras for the year 1930-31, 20 pp.

369. Sundaraman, S. 1936. Administration Report of the Mycologist, Madras for the year 1935-36, 13 pp.

370. Sung. J.M., Park J.H., Lee S.C. and Chung B.K. 1980. The out break and propagule formation of black root rot caused by *Calonectria crotalariae* in Korea. *Korean Journal of Plant Protection* 19 : 228-233.

371. Tai, F.L. 1936. Notes on Chinese fungi VI. *Bulletin Chinese Botanical Society* 2 (1) : 16-18.

372. Tai, F.L. 1936. Note on Chinese fungi VII. *Bulletin Chinese Botanical Society* 2 (2) : 45-66.

373. Tandon, R.N. 1976. *Achaetomium* : A new record of seed borne pathogen of *Phaseolus aureus. Indian Phytopathology* 29 : 464-465.

374. Taneja, M. and Grover R.K. 1982. Efficacy of benzimidazole and related fungicides against *Rhizoctonia solani* and *Rhizoctonia bataticola. Annals of Applied Biology* 100 : 425-432.

375. Tarr, S.A.J. 1954. Plant Pathology. *Report Research Division Ministry of Agriculture, Sudan 1951-52,* 71-80 pp.

376. Thakur, M.P. 1988. Influence of the environmental conditions on the incidence of anthracnose of mungbean. *Indian Phytopathology* 41 : 281.

377. Thakur, M.P. and Khare M.N. 1989. Evaluation of mungbean varieties for resistance to anthracnose. *Indian Journal of Plant protection* 17 : 107-108.

378. Thakur, M.P. and Khare M.N. 1989. *In vitro* testing of different fungicides against two species of *colletotrichum* causing anthracnose of mungbean. *National Symposium* on *New Frontiers of Pulses Research and Development, Kanpur* : 114 p.

379. Thakur, R.P., Patel P.N. and Verma J.P. 1977. Independent assortment of pigmentation and resistance to *Cercospora* leaf spot in mungbean. *Indian Phytopathology* 30 : 264-265.

380. Thakur, R.P., Patel P.N. and Verma J.P. 1980. Inheritance of resistance to Cercospora leaf spot in mungbean. *Indian Phytopathology* 33 : 377-379.

381. Thind, B.S. and Kumar Ish. 1980. Chemical control of bacterial and *Cercospora* leaf spots and leaf reddening of mungbean. *Indian Journal of Mycology and Plant Pathology* 10: LXXVIII and LXXIX.

382. Thirumalachar, M.J. and Mishra J,N. 1953. Some diseases of economic plants in Bihar, India. *FAO Plant Protection Bulletin* 10 : 145-146.

383. Tiwari, Anamika and Kotasthane S.R. 1984. Chemical control of fungal foliar diseases of moong. *Indian Phytopathology* 37 : 383.

384. Tiwari, Anamika and Kotasthane S.R. 1986. Chemical control of fungal foliar diseases of mungbean. *Pesticides* 20(12) : 47-48.

385. Tiwari, A.S. 1977. Mungbean varieties requirements in relation to cropping seasons in India. *First International Mungbean Symposium, AVRDC, Taiwan,* 129-131 pp.

386. Tripathi, H.S. and Beniwal S.P.S. 1977. Anthracnose of urd bean caused by *Colletotrichum capsici. Indian Phytopathology* 30 : 273-274.

387. Tseng, T.C. and Chang L.H. 1974. *Sclerotium rolfsii* phosphatidase BI inhibition of mungbean mitochondrial oxygen uptake by phosphatidase. *Botanical Bulletin of Academia Sinica* 15 : 8-13.

388. Tsiung, Ng Thai 1977. Response of mungbean to sowing dates in Sarawak. *First International Mungbean Symposium, AVRDC, Taiwan,* 101-106 pp.

389. Tsuchiya, S., Yanagawa W. and Ogoshi A. 1986. Formae speciales differentiation of *Phytophthora vignae* isolates from cowpea and adzuki bean. *Annals of the Phytopathological Society of Japan* 52 : 577-584.

390. Tyagi, R.N.S., Mathur A.K., Gaur V.K., Chitley K., Bansal R.K. and Pathak A.K. 1988. Pathological status of pulse crops in Rajasthan. *Indian Phytopathology* 41: 280.

391. Uozumi, T. and Yoshii H. 1953. Some observations on the mildew fungus affecting the cucurbitaceous plants. *Annals of the Phytopathological Society of Japan* 16 : 123-126.

392. Uppal, B.N., Patel M.K. and Kamat M.N. 1935. The fungi of Bombay 8 : 1-56.

393. Utikar, P.G., Deshmukh R.B. and More B.B. 1978. Field reaction of blackgram (*Vigna mungo* (L.) Hepper) germplasm to powdery mildew. *Tropical Grain Legume Bulletin* 13/14 : 30-31.

394. Vaheeduddin, S. 1955. Phytopathological survey of Hyderabad State. *Indian Phytopathology* 8: 166-171.

395. Vasudeva, R.S. 1961. "Indian Cercosporae". *Indian Council of Agricultural Research, New Delhi,* 245 pp.

396. Verma, P.R. and patel P.N. 1969. Host range, varietal resistance and eipidemiological observations on *Cercospora* leaf spot disease of cowpea. *Indian Phtopathology* 22 : 61-66.

397. Vidhyasekaran, P. and Kandaswamy D. 1972. Carbohydrate metabolism of *Phaseolus aureus* infected with obligate, facultative parasites. *Indian Phytopathology* 25 : 48-54.

398. Vidhyasekaran, P. and Arjunan G. 1976. Resistance in some urd varieties against powdery mildew. *Indian Phytopathology* 29 : 212-213.

399. Vidhyasekaran, P. and Arjunan G. 1978. Studies on leaf blight of urdbean caused by *Macrophomina phaseoli. Indian Phytopathology* 31: 361-362.

400. Vidhyasekaran, P., Arjunan G. and Ranganathan K. 1976. Field tolerance of some blackgram varieties to root rot disease caused by *Macrophomina phaseolina. Madras Agricultural Journal* 63 : 176-178.

401. Vidhyasekarn, P., Arjunan G. and Ranganathan K. 1977. Resistance in some mung varieties against root rot. *Indian Phytopathology* 30 : 404.

402. Vignarajah, V. 1977. Mungbean research and production in Sri Lanka. *First International Mungbean Symposium, AVRDC, Taiwan* : 9-11 pp.

403. Vinayak, M.S. 1977. Field evaluation of fungicides and cultural methods to control powdery mildew (*Erysiphe polygoni*) of greengram (*Phaseolus aureus* Roxb.). M.Sc. Ag. Thesis, M.Phule Krishi Vidyapeeth, Rahuri, India, 89 pp.

404. Wang, T.C. and Yang C.Y. 1976. Mungbean leaf blight, a new disease of mungbean. *Plant Protection Bulletin* 17(4) : 8-9.

405. Welles, C.G. 1924. Studies on leaf spots of *Phaseolus aureus* new to the Philippines islands. *Phytopathology* 14 : 351-358.

406. Williams, F.J., Grewal J.S. and Amin K.S. 1968. Serious and new diseases of pulse crops in India in 1967. *Plant Disease Reporter* 52: 300 304.

407. Wilson, K.I. and Jose P.C. 1968. Pod rot of *Vigna sinensis* (L.) Savi caused by *Choanophora cucurbitarum* (Berk. Rav.) Thaxter. *Indian Phytopathology* 31 : 130-131.

408. Wu, L.C. 1965. Physiology of parasitism. I. Growth, pathogenicity and toxin production of *Rhizoctonia solani* Kuhn. *Botanical Bulletin of Academia Sinica* 6 : 144-152.

409. Wu, L.C. 1967. Physiology of parasitism. II. Biochemical changes in the mungbean seedlings infected with *Rhizoctonia solani* Kuhn, *Botanical Bulletin of Academia Sinica* 8 : 271-283.

410. Wu, L.C. 1969. Physiology of parasitism. III. Nitrogen metabolism in mungbean seedlings infected with *Rhizoctonia solani. Boatanical Bulletin of Academia Sinica* 10 : 95-108.

411. Wu, L.C. 1969. Physiology of parasitism. IV. *Rhizoctonia* metabolites, amino acids and others. *Botanical Bulletin of Academia Sinica* 10 : 109-123.

412. Wu, L.C. 1973. Changes in some enzymes of mungbean seeds germinated on mycelial macerates of *Rhizoctonia solani. Physiological Plant Pathology* 3 : 19-27.

413. Yang, C.Y. 1977. Mung bean diseases and control. First *International Mungbean Symposium AVRDC, Taiwan* 141 - 146 pp.

414. Zainum WAN Nik, and Ngehang J. 1982. Studies of *Rhizoctonia solani* Kuhn on mungbean (*Vigna radiata* (L.) Wilczek). *Malaysian Applied Biology* 11 : 103- 110.

415. Zaki, M.J. and Ghaffar A. 1987. Effect of *Rhizobium* spp. on *Macrophomina phaseolina. Pakistan Journal of Scientific and Industrial Research* 30 : 305-306.

416. Zapromotov, N.G. 1928. Work done in 1926 by the Phytopathological Station, report for 1926. *La Defense des Plant Leningrad* v, 1:75-79.

417. Zapromotov, N.G. and Gafurov R.K. 1972. Control of infectious diseases of oregan pea. *Zashchita Rastenii* 7 : 27.

418. Zote, K.K. Dandnaik B.P. and Jature S.D. 1981. Chemical control of foliar diseases of mungbean *(Vigna radiata)*. *Pulse Crops Newsletter* 1 (2) : 68.

419. Zote, K.K., Dandnaik B.P. and Khalikar P.V. 1983. Reaction of mung cultivars to *Macrophomina* blight. *Journal of Maharashtra Agricultural Universities* 8 : 146 - 147.

420. Zote, K.K., Khalikar P.V. and Dandnaik B.P. 1985. Efficacy of fungicides against powdery mildew of mung. *Pesticides* 19(8): 42-43.

ADDENDA

421. Abbaiah, K. and Satyanarayana S. 1990. Stem canker of black gram (*Vigna mungo* (L.)Hepper) in Andhra Pradesh. *Indian Journal of Plant Protection* 18 : 136-137.

422. Agrawal, S.C. and Nema Sushma. 1984. Plant Pathology. In Annual Report of All India Coordinated Pulses Improvement Project, Sehore, M.P., India, 1983-1984 : 30-41 pp.

423. Agrawal, V.K., Mathur S.B. and Neergaard P. 1972. Some aspects of seed health testing with respect to seed borne fungi of rice, wheat, blackgram, greengram and soybean grown in India. *Indian Phytopathology* 25 : 91-100.

424. Ahmad, H.U., Bakr M.A., Alam K.B. and Kaul A.K. 1981. Diseases of Major pulse crops in Bangladesh. *III International Symposium Plant Pathology, IPS, New Delhi*, 86 pp.

425. Arx, J.A.V. 1957. Die Arten der Gattung *Colletotrichum* Corda. *Phytopathologische Zeitschrift* 29 : 413-468.

426. Chaubey, U. 1978. Studies on seed borne fungi of mung with special reference to *Fusarium* spp. M.Sc. Ag. Thesis, J.N.K.V.V.. Jabalpur, M.P., India, 96 pp.

427. Chaubey, U., Khare M.N. and Agrawal S.C. 1978. Efficacy of fungicides in the control of *Fusarium oxysporum* associated with mung bean seeds. *Pestology* 4 (4) : 32-34.

428. Chindhalore, J.L. 1974. Studies on the role of seed borne pathogens of mung causing seed and seedling diseases and their control. M.Sc.Ag. thesis, J.N.K.V.V., Jabalpur, M.P., India, 58 pp.

429. Jharia, H.K. 1970. Studies on seed pathology of pulse crops important in Madhya Pradesh. M.Sc.Ag. Thesis, J.N.K.V.V., Jabalpur, M.P., India, 87 pp.

430. Jhooty, J.S. and Bains S.S. 1976. Modification of the toxicity of benomyl by seed exudates of mung. *Indian Journal of Microbiology* 16: 73-75.

431. Kaiser. W.J., Danesh D., Okhovat M. and Mossahebi H. 1968. Diseases of Pulse crops (edible legumes) in Iran. *Plant Disease Reporter* 52 : 687-691.

432. Kanta, S. and Vaidehi B.K. 1980. Effect of culture filtrate on the seed germination and on root and shoot elongation of *Phaseolus mungo*. *Indian Journal of Mycology and Plant Pathology* 10: XXII.

433. Khare, M.N., Mathur S.B. and Neergaard P. 1977. Seed borne fungi of pulses, their detection, location and importance. *XVIII International Seed Testing Association Congress, Medrid, Spain, Reprint No. 80-S* : 3 pp.

434. Kulik, M.M. 1984. Symptomatology and epidemiology of several green bean diseases incited by seed borne fungi. *Seed Science and Technology* 12 : 841-850.

435. Kulshreshtha, D.D., Mathur S.B. and Neergaard P. 1976. Identification of seed borne species of *Colletotrichum*. *Friesia* 11 : 116-125.

436. Latham, D.H. 1934. Life history of a *Cercospora* leaf spot fungus of cowpea. *Mycologia* 26 : 516-527.

437. Mohanty, U.N. and Rao V.B. 1968. In Proceedings II Annual Workshop Conference on Pulses, IARI, New Delhi.

438. Moore, E.S. 1931. Internal boll disease of cotton in South Africa. *South Africa Department of Agriculture Science Bulletin* 94 : 11-18.

439. Nath, R., Mathur S.B. and Neergaard P. 1970. Seed borne fungi of mung (*Phaseolus aureus* Roxb.) from India and their significance. *Proceedings International Seed Testing Association* 35 :225-241.

440. Rath G.C. and Routray G.N. 1978. Internally seed transmitted diseases of mung and urid. *Science and Culture* 44 : 40-41.

441. Rayan, G.F., Greenblatt G. and Dolaimy K.A.A. 1961. Seedling albinism induced by an extract of *Alternaria tenuis*. *Science* 134 : 833-834.

442. Sandhu, T.S. 1977. Breeding for yellow mosaic resistance in mungbean. *I International Mungbean Symposium, AVRDC, Taiwan* 176-179 pp.

443. Saxena, R.M. and Sinha S. 1979. Field mycoflora of *Vigna radiata* (L.) Wilczek var. *radiata* and *Vigna mungo* (L.) Hepper in relation to pre emergence and post emergence mortality. *Seed Research* 7 : 159-164.

444. Shukla, D.N. and Bhargava S.N. 1976. Some pathogenic fungi from pulses and oil seed crops. *Proceedings National Academy of Sciences India (B)* 46 : 531-532.

445. Singh, D.P. and Sharma B.L. 1981. Evaluation of mungbean germplasm. *Madras Agricultural Journal* 68 : 289-295.

446. Singh. Leena and Srivastava S.N. 1985. Studies on morphology and taxonomy of *Colletotrichum* species parasitizing pulse crops in Northern India. *Indian Phytopathology* 38 : 616.

447. Singh, P.P. and Thind B.S. 1981. Histological Changes and enzyme production in mungbean in response to infection with *Xanthomonas phaseoli f. vignaeradiatae* and *Cercosopora canescens. III International Symposium* of *Plant Pathology, IPS, New Delhi,* 202-203 pp.

448. Singh, P.P. and Thind B.S. 1984. Interaction between *Xanthomonas campestris* pv. *vignaeradiatae* (Sabet et al.) Dye and *Cercospora canescens* Ellis et Mart. on mungbean in relation to disease development. *Phytopathologia Mediterranea* 23 : 75-76.

449. Singh, P.P. and Thind B.S. 1987. Histological changes in green gram leaves in response to *Xanthomonas campestris* pv. *vignaeradiatae* and *Cercospora canescens* infection. *Indian Phytopathology* 40 : 149-151.

450. Singh, R.R. and Shukla P. 1988. *Colletotrichum* leaf blight of black gram. *Farm Science Journal* 3 : 97-102.

451. Talbot, P.H.B. 1970. `Principles of Fungal Taxonomy'. McMillan London.

452. Thind, B.S. and Kumar Krishan. 1985. Chemical control of bacterial leaf spot and *Cercospora* leaf spot of mungbean. *Indian Phytopathology* 38 : 601.

453. Verma, M.L. 1985. Control of foliar diseases of urd bean (*Vigna radiata* var. *mungo*). *Indian Phytopathology* 38 : 600-601.

454. Vestal, E.F. 1956. "A text Book of Plant Pathology". Allahabad Kitabistan.

455. Walker, J.C. 1969. "Plant Pathology". 3rd Ed. McGraw Hill Inc. New York, 819 pp.

456. Young, P.A. 1949. Charcoal rot of plants in East Texas. *Texas Agricultural Experiment Station Bulletin No. 712.*

CHAPTER 3
BACTERIAL DISEASES

Blackgram and greengram suffer from leaf spot and halo blight caused by bacteria. Most of the work on these diseases has been done after 1963 only. The diseases have been reviewed recently (10, 20).

3.1. BACTERIAL LEAF SPOT :

The first report of the bacterial leaf spot of greengram seems to be from Bambari (4) in 1960. However, the first published record of disease occurrence is from China by Fang *et al.* (12). Later, the disease was recorded from Sudan (36) and India (30). In India, it has been found occurring in the states of Delhi, Uttar Pradesh, Punjab, Haryana, Madhya Pradesh, Tamil Nadu, Karnataka, Rajasthan, Maharashtra and Himachal Pradesh (20, 23,31,41,53). The disease is wide spread wherever the crop is grown (CMI map No. 401). On blackgram, the disease was not described earlier (34). It is indicated that the bacteria causing leaf spot on greengram and blackgram are different from each other (32).

Symptoms :

On greengram, the symptoms are characterized by the presence of brown, circular to irregular and distinctly raised spots on leaves (Fig. 3.1) and have been studied in detail (32). The spots are dry and necrotic from the beginning, sometimes water-soaked or with translucent border. They appear on both the surfaces of leaf as superficial eruptions but later develop through the entire thickness of the leaflet and become corky or rough. After attaining some height, the eruptions begin to collapse due to disintegration of the infected tissues resulting in rifts. In some cases, corky cells wither and spots appear sunken or a saucer like. Closely situated spots may coalesce forming big necrotic areas but typical leaf blight phase is not developed as reported in case of beans. The lesions on petiole, petiolute or pulvinus are brown, flat or raised occurring singly or in the form of long streaks. The disease does not cause defoliation. On stems also similar spots are produced (Fig. 3.2) but in cases of severe infections dark brown longitudinal slits or cankers develop which often extend from soil level to the growing tip. Lesions on pods are raised and brown but seen rarely in nature. Variation in symptoms have been noted on greengram variety CO 1 which shows appearance of minute water-soaked lesions in the beginning but later they enlarge in size leading to blight upon coalescing and after surrounded by mild halo (23). These variations in symptoms are explained to be due to existence of different strains of the pathogen.

Fig.3.1.
Bacterial leaf spot of green gram (Courtesy : Dr. K. Abbaiah)

On blackgram, the symptoms develop in 5-8 days after inoculation as circular to irregular water soaked areas which later develop in to distinctly raised spots and turn dark brown in colour (22). Spots are dry, necrotic and coalesce to form necrotic areas but typical leaf blight phase is not developed. Similar spots develop on stem but extensive pod infection is rare (20).

Causal organism :

Xanthomonas campestris pv. *vignae-radiatae* (Sabet et al.) Dye on greengram.

X. compestris pv. *phaseoli* (Smith) Dye Urdbean strain on blackgram.

The pathogen causing leaf spot was identified as *X. phaseoli* (Smith) Dowson on greengram (12) as well as blackgram (34). Though, the blackgram and greengram strains of the pathogen are not much different in their bacteriological characters but they differ much in their host range, and pathogenicity. On the basis of new standard of nomenclature (11), the greengram strain is named as *X. campestris* pv. *vignae-radiatae* while blackgram strain is considered as a strain of *X. campestris* pv. *phaseoli*.

Cultural and morphological characters : The colonies of the greengram pathogen on nutrient agar are circular, convex, wet, shining and yellow in colour. Morphologically, the bacterium is characterized by gram-negative rods of 0.87 x 1.9μ size with monotrichous flagella and forms slime (32). The blackgram strain has typical gram negative rods of 0.5 to 0.8 x 1.2 to 1.7 μ size with monotrichous flagella. It is non capsulated, non spore former and grows well in nutrient agar and mannitol agar media (22).

Physiological and biochemical characters :According to Patel and Jindal (32) the greengram isolate is strictly aerobic; NaCl toleration 3 %; utlilization of glucose oxidative; does not utilize asparagine; peptonized action in milk without curd; hydrolyses gelatin and starch; produces ammonia and catalase but not indole, acetone and urease; does not accumulate NO_2 from NO_3; forms H_2S from cysteine; produces acid from mannitol, mannose, glucose, sucrose, fructose, galactose, ribose, arabinose, melibiose, cellobiose, trehalose, lactose, xylose, raffinose, dextrin, glycogen but not from sorbitol, rhamnose, adonitol, dulcitol, inulin, sorbose, salicin, inositol and mesoerythritol; utilizes sodium propionate and sodium succinate but not sodium oxalate and sodium tartarate.

The characters of the blackgram pathogen resemble with those described for greengram isolate (22). It is aerobic; liquifies gelatin; peptonizes litmus milk; hydrolyses starch; produces H_2S and ammonia but not indole; does not reduce nitrate and give methyl red test positive. It grows well at pH 7.0 and 29°C temperature and has a thermal death point at 51°C. The bacterium utilizes xylose, saccharose, lactose, maltose, galactose,mannose, raffinose, mannitol, dextrose, fructose and trehalose but not rhamnose and salicin. Amongst the nitrogen compounds, ammonium nitrate, ammonium sulphate, ammonium chloride, potassium nitrate, sodium nitrate, calcium nitrate, glycine and urea are utilized but not barium nitrate, magnesium nitrate and asparagine.

Variability in the pathogen : Although, there may be overlapping in host range and similarity in cultural, physiological and biochemical characters but isolates of *Xanthomonas* from different leguminous plants are more virulent on their natural hosts and hence, each isolate should be considered as a strain or race of the pathogen (12). It is very difficult to differentiate the isolates on the basis of bacteriological properties but they vary in colony size, mucoidness and pigmentation. Mucoidness is considered to be related with the degree of virulence. The white mutants of mungbean isolates have been found to be as virulent as their yellow forms. As compared to mucoidness, rate of multiplication is not much important in virulence (18). Seven biochemical mutants of mung bean *Xanthomonas* have been isolated which are either avirulent or less virulent on their host (58).

Fig.3.2.
Spots on green gram stem caused by *Xanthomonas campestris* pv. *vignae-radiatae*.

Not only the isolates from different host plants differ from each other but even isolates obtained from the same host species differ in their pathogenic behaviour and production of symptoms (23). In India, six distinctly pathogenic races of greengram pathogen have been identified by using 4 differentials (16,56). The differentials are P 23, P 10875, PLM 95 and IC 11303-3.

Several methods have been tried to differentiate the strains / isolates of *Xanthomonads* of pulses in the lab. It is reported that the isolates of *X. campestris* pv. *phaseoli* from different hosts differ in their soluble protein pattern (SSP) but there is no correlation between the SSP of an isolate and its geographic distribution, virulence, attackability, colony variation and the phage sensitivity (3). Jindal and Patel (16) tried to explore the possibility of identifying the greengram strains of *Xanthomonas* by using bacteriophages but they were not successful.

Host range : The greengram and the blackgram isolates of *Xanthomonas* have been studied in detail in regard of their host range and both have been found to differ from each other to a greater extent (Table 3.1).

Table 3.1

Pathogenicity of greengram and blackgram isolates of *Xanthomonas campestris*

Host	Pathogenic (+) or non pathogenic (-)	
	Greengram isolate Patel & Jindal(32)	Blackgram isolate Kore & Shirshikar(22)
Phaseolus aureus Roxb.	+ + + +	-
P. vulgaris L.	+ + +	+ +
P. bracteatus Ness & Mart	+	
P. mungo L.	-	+ + +
P. lunatus Benth	+	+ +
Vigna sinensis (Torner Savi)	-	-
Dolichos lab lab L.	+ +	+ +
D. biflorus L.	-	+ +
Cyamopsis tetragonoloba (L.) Taub.	-	-
Taphrosia purpurea Pers.	-	
Lens esculentum Moench.	+ +	+ +

It is clear that the two isolates were highly pathogenic to their specific hosts. The greengram isolate was non pathogenic on blackgram (*P. mungo*) and the blackgram isolate did not infect greengram (*P. aureus*). On other hosts also, they had different pathogenic reactions. However, Rangaswamy and Gowda (34) isolated a strain of *Xanthomonas campestris* pv. *phaseoli* and reported it pathogenic to blackgram as well as greengram. It caused leaf spot of blackgram in Madras state. It infected *Cyamopsis psoalioides, D. lablab, P. vulgaris, Vigna catjang* and *Taphrosia purpurea* also which were not infected by the greengram isolate from Delhi (32). It indicates that the isolate from Madras (34) was different from that of blackgram isolate (22) and greengram isolate (32) studied previously. It may be the bean common blight pathogen which is reported to infect blackgram and greengram both (8, 63).

Histopathology and host parasite interaction : The bacterium is mainly parenchymatous and infection of greengram leaves causes certain morphological changes (46, 48). It causes rupturing of epidermal, palisade and spongy parenchymatous cells; increases intercellular spaces; degradation of cell walls; collapse of cells and disorganization and scattering of cytoplasmic contents. Infection in leaves results in bacterial pocket formation of wounds, periderm or cicatrice separating the diseased portion from healthy and attack on vascular bundles.

When greengram plants are rub inoculated by the leaf spot bacterium, different types of reactions are produced according to their degree of resistance (33). In suspectible plants, whitish, blister like spots develop which later show rift in them and turn necrotic. In tolerant plants, brown, necrotic and occasionally sunken, minute spots develop within 2-3 days suggesting a kind of hypersensitive reaction which is not detected in spray inoculations of same variety. The resistant lines give strong hypersensitive reaction which can be used as a criteria for selecting a resistant variety.

The pathogen produces a host non specific toxin which inhibits many fungi and causes necrotic brown spots on greengram as well as other non host plants (25). This toxin might be playing some role in producing disease symptoms. Pathogen produces less toxin in moderately resistant (CO 2) as compared to susceptible varieties (CO 1 & CO 3). This difference has been attributed to the presence of higher quantity of cysteines, a sulphur containing amino acid, in resistant as compared to susceptible varieties (27). The spray of 0.05 % cysteine on susceptible variety reduces the disease intensity to a greater extent.

The permeability of the cell membrane of host is increased which increases leakage of ions, amino acids and sugars. Such type of effect is more in susceptible as compared to resistant variety (24).

The infection increases the activity of polyphenoloxidase and peroxidase. The resistance in greengram variety CO 2 is related to increased peroxidase activity and ultimately increase in total phenol contents of the infected tissues (29).

The nitrogen metabolism of the infected greengram plant is also affected. Initially, nitrogen content of a susceptible variety (Jawahar 45) is higher as compared to resistant vrieties (ML 8 & Jalgaon 781) but after infection the nitrogen as well as free amino acids decrease to a greater extent. On the other hand Marimuthu and Kandaswamy (28) reported that a moderately resistant variety of greengram (CO 2) contains more amino nitrogen than the susceptible ones (CO 1 & CO 3). The infection increases nitrogen content of CO 2 but decreases in CO 1 and CO 3. In fact, the proportion of aromatic amino acids in the amino nitrogen pool and the presence of sulphur containing amino acids are related with the resistance of CO 2. These compounds inhibit the growth of the pathogen and production of Endo-PG and PGTE by it.

Sugar contents of host tissues also contribute to resistance of host plant. The greengram variety CO 1 which is susceptible to the pathogen, possesses high level of reducing sugar while resistance in CO 2 is due to low glucose content (26). The total sugars, total phenols and protein contents of greengram variety inoculated with *X. campestris* pv. *vignae-radiatae* has been found to be higher as compared to that infected by *Cercospora* sp. (47).

Disease cycle and epidemiology :

Field observations suggest that the pathogen in greengram is seed borne through which the dissemination takes place in the new area while local spread of the pathogen is through spattering rains. The incidence of the leaf spot disease in the field has been found correlated with the percentage of seed infected by the pathogen (42). After the secondary transmission, the pathogen enters the host tissues through stomata and the infection is favoured by the injury given to the host. It facilitates the infection and decreases the incubation period also (5, 22). The advancing age of the plant increases resistance in blackgram agaisnt its pathogen. The plants inoculated at the age of 15 days showed 70% infection but no infection was observed in plants at the age of 65 days (22). The optimum bacterial population level for the disease development in greengram has been reported to be 3.5 x 10^3 cells per ml of suspension (5).

Environmental factors greatly influence the disease development and spread. The disease on greengram is reported less severe during summer with a mean RH 20-80 % and temperature 25-34°C, but it increases during the period of monsoon when mean RH is 50-95 % and the temperature ranges between 24-32°C (42). It has been found that continuous heavy rains coupled with low temperature (25-30°C) and high RH (80%) favour the disease severity and spread (2). The wind velocity has been found negatively correlated with the disease progress on susceptible variety LBG 403.

Host resistance :

Sources of resistance have been identified against the bacterial leaf spot of greengram as well as blackgram (Table 3.2). The inheritance of resistance in mungbean lines (P476, P 646, PLM 501 and Jalgaon 781) is governed by a single dominant gene which is inherited and combined independent to seed colour and maturity (44). On the other hand, resistance in greengram varieties ML 8, ML 10 and Jalgaon 781 is reported to be governed by a single recessive gene (62). The genetic make up of the four greengram differentials P 23, P 10875, PLM 95 and IC 11303-3) has been studied using six races of the pathogen with the following conclusions (56, 57).

1. Resistance in a differential to a race is governed by a single gene.
2. Gene governing resistance in a differential is dominant to its recessive allele governing susceptibility in another differential.
3. The gene or allele governing resistance or susceptibility to different races do not overlap.

Table 3.2
Varieties/genotypes of greengram and blackgram reported resistant against bacterial leaf spot

Host species	Entries showing resistance	Reference
Vigna radiata (greengram)	Out of 23 entries Jalgaon 781 was highly resistant	(32)
	Out of 2160 germplasm types, P 56, P 161, 200, 275, 277, 278, 441, 445, 476, 530, 646, 652, PLM 5, 16, 19, 90, 95, 123, 147, 161, 243, 249, 439, 496, 501, IC 11303-3, EC 13077, 19743 and Jalgaon 781 were resistant.	(33)
	Out of several entries, LM 113, 168, 170, 171 and 404 were resistant.	(38)
	Out of 90 entries,, ML 8 ML 10 and Jalgaon 781 were highly resistant and ML 1, ML 2, ML3, ML 9, Yellow mung and P 70-68 were resistant.	(9)
	Out of 25 cvs., CO 2, K 851, PLS 292, 278 and 266 were moderately resistant	(23)
	Out of 4380 germplams lines, ML 234, 268, 269, 272, 274, 338, 360, 369, 371, LM 113, 687, 695, 696, 744, MUG 125, 130, 132 to 136, 138, 139, 145, 170, 173, M 13 and Tarai local were resistant.	(45)
	Out of several entries, Mutant of Pant Mung 2 was highly resistant and 11 lines including ML 5 were MR	(1)
Vigna mungo (blackgram)	Out of 22 lines, Krishna was HR, H 70-3, US 131, Urd 65, G 104, T 9 and Urd H 10 were R	(22)
	Out of 196 entries, 6 were tolerant	(15)
	Out of 12 cvs., JU 78-16, JU 78-27, JU 78-52 and Plant U 30 were resistant	(61)

It has been reported that the resistant genotypes of greengram produce hypersensitive reaction against *X.campestris* pv. *vignae-radiatae* which can be used as a criteria for selecting resistant plants during screening for disease resistance (33).

Control :

Use of disease free seed : Since, the disease is seed transmitted (32), procuring disease free seed is of prime importance. The disease severity during summer remains at its minimum level. Hence, disease free seed can easily be produced by growing the seed crop in summer and carefully roguing of infected seedlings (42). Summer period is not suitable for detecting the presence of *X. campestris* pv. *vignae-radiatae* in seed of greengram by growing on test. For this purpose monsoon period is suitable (52).

Seed treatment by heat : The pathogen in greengram seed can be inactivated by hot water treatment at 52°C for 20 minutes (14). Solar heat treatment or hot water treatment at 52°C or 30 minutes has also been found effective (59, 60).

Chemotherapy : The greengram strain of the pathogen is inhibited by chloramphenicol or vitavax (6) however, results *in vivo* are unlike of those obtained *in vitro*. It is inhibited completely by 500-1000 ppm stable bleaching powder *in vitro* test (37). Seed treatment with emisan-6 and spray of crop with stable bleaching powder (500-1000 ppm) control the disease effectively increasing yield

For seed treatment, captan (0.3%) and bleaching powder (0:025%) have been found to be most effective (53, 54). In field experiment, 3 protective sprays of each of streptocycline (100 ppm), zineb (0.3%) or benomyl (0.2%) proved effective in controlling the secondary infection by the bacteria.

On the basis of 3 years data, seed treatment with agrimycin 100 (700 ppm) followed by three sprays of agrimycin 100 (0.025%) + bavistin (0.1%) have been found very effective for control of bacterial leaf spot of greengram (13, 14).

Thind and Kumar (60) recommended a schedule of fungicidal use for the control of leaf spots of greengram caused by bacterium and *Cercospora* spp. It includes seed treatment with streptocyline

(100 ppm) + agallol (0.2%) or streptocyline (100 ppm) + captan (0.2%) coupled with any of the following schedule :

1. First 2 sprays of streptocyline (100 ppm) + bavistin (0.05%) followed by last 2 sprays of streptocyline (100 ppm) + blitox 50 (0.2%).
2. First 2 sprays of streptocycline (100 ppm) + bavistin (0.025%) + blitox 50 (0.1%) followed by last 2 sprays of blitox 50 (0.2%).

Verma on the other hand reported bavistin and benlate to enhance the severity of bacterial leaf spot of blackgram and found copper oxychloride to be effective (61).

The seed treatment with hot water, solar heat, streptocycline, captan or streptocycline + agallol reduces diseased seedlings as well as intensity of bacterial leaf spot of greengram. In addition, spray with fungicide/antibiotics reduces bacterial as well as Cercospora leaf spots and increases the yield (19). However, it is essential to find out the economics of treatment before recommending any fungicidal schedule for a crop like greengram.

Biological control : During *in vitro* evaluation, several bacteria and fungi including *Bacillus cereus, B. subtilis, Erwinia herbicola, Enterobactor aerogens, Pseudomonas fluorescens, Epicoccum nigrum, Penicillium oxalicum, Trichoderma viride,* were found antagonistic against *X.* campe-stris pv. *vignaeradiatae* and the antagonistic metabolites produced by them depended on the medium, its pH and the temperature of incubation (51). These antagonists when applied to greengram seed, significantly reduced the disease incidence in green house as well as in field and increased the yield up to 34.4%. *B. cereus* was highly effective in this respect followed by *B. subtilis*. Both these bacteria persisted in the rhizosphere while *E. aerogens* and *P. fluorescens* persisted in the phyllosphere of greengram (49).

3.2. HALO BLIGHT OF GREENGRAM :

This diseas was first observed at Delhi, India or some greengram lines resistant to bacterial leaf spot in the year 1969 (30). The disease was indicated to be caused by a Pseudomonad which did not infect the bean and soybean (32). A severe epiphytotie of the disease was reported from U.S.A. in 1971 (39). It is reported to occur widely in the northern parts of Tanzania (20). In India, the disease has been reported from the state of Maharashtra also (43). The distribution of the pathogen is given in the CMI distribution map No. 85.

Symptoms :

The symptoms of the disease have been described in detail (32). Water-soaked spots appear on cotyledonary leaves of seedligns produced from the infected seed. Heavy infection may cause death of seedling. The pathogen later spreads to the first trifoliate and other leaves forming small or large, round or irregular, flat, water-soaked spots surrounded by a greenish-yellow zone (halo). The halo is produced probably due to the toxins secreted by the pathogen in the water-soaked area. On a heavily infected plant, the uninoculated leaves also become chlorotic due to translocation of toxin. The halo formation and the systemic chlorosis is favoured at low temperature (15-20°C) at night. Spots on old leaves are sometimes angular without any distinct halo (Fig. 3.3). Frequently the infection spreads along the veinlets and veins turning them brown and necrotic and causing distortion of the developing leaflets. On petioles, long water-soaked streaks are formed which later turn red. Pulvinus and petiolutes of the severely infected leaf appear brown and sunken. Plants do not defoliate, though heavy leaf shedding due to the disease is reported in Maharashtra (43). On the stem, the lesions are water-soaked which gradually enlarge developing in to reddish streaks which extend longitudinally along the stem and often developing splits (Fig. 3.4). On pods, lesions are water-soaked which later turn brown.

Causal organism :

Pseudomonas syringae pv. *phaseolicola* (Burk.) Young, Dye and Wilkie mungbean strain.

The pathogen seems to be isolated first by D.C. Graham from greengram seeds in Scotland in the year 1966 who identified the culture as *P. phaseolicola* (40). This culture was later studied by Schroth *et al.* (40) who found it to be different from the bean isolate of *P. phaseolicola*. Hence, Schmitthenner *et al.* proposed it as a distinct strain of *P. phaseolicola* which has been renamed as

P. syringae pv. *phaseolicola* on the basis of new standards of nomenclature (11).

Cultural and morphological charaters : On nutrient agar, the colonies of the bacterium are circular, flat, entire, glistening, white and produce green fluorescent pigment in King's medium. The bacterium is a gram negative rod with 1-4 polar flagella without any slime formation (32).

Fig.3.3.
Halo blight symptoms on old leaf of green gram.

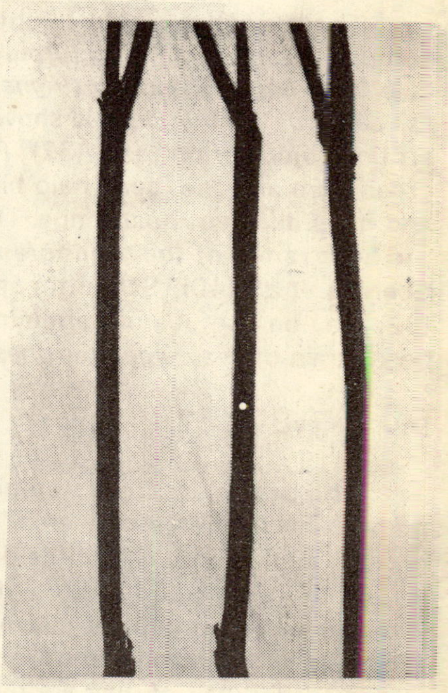

Fig.3.4.
Green gram stem infected by *Pseudomonas syringe* pv. *phaseolicola*.

Physiological and biochemical characters : The mung bean halo blight bacterium differs from the bean halo blight strain (32). It is strictly aerobic with NaCl tolerance 4%; oxidative glucose utilization and peptonization of milk without curd formation. It hydrolyses gelatin (slowly) and starch; utilizes asparagine, sodium propionate, sodium succinate but not sodium oxalate and sodium tartarate. It does not cause accumulation of NO_2 from NO_3 and does not produce H_2S from cysteine. It produces ammonia urease and catalase but not indole and acetone. The bacterium produces acid from mannose, mannitol, glucose, sucrose, fructose, galactose, ribose, arabinose, melibiose, cellobiose, lactose, xylose, raffinose but not from trehalose, sorbitol, rhamnose, adonitol, dulcitol, inulin, sorbose, salicin, inositol, meso erythrite, dextrin and glycogen.

Host range :

The mungbean halo blight pathogen is most virulent on *P. aureus (V. radiata), P. atropurpureus, P. bracteatus, Dolichos lablab, D. biflorus while P. mungo (Vigna mungo), P. trilobus, V. sesquipedalis, Lathyrus sativus* do not show uniformely positive reaction under artificial inoculation tests (32). *Ficus religiosa* is also infected. The greengram and the bean halo blight pathogens do not cross-infect and have different host range. The detailed comparative studies on the host range of the greengram halo blight pathogen isolated by Schroth *et al.* (40), Schmitthenner *et al.* (39) and Patel and Jindal (32) may be useful in determining the extent of variation in mung bean strain of *P. syringae* pv. *phaseolicola*.

Disease cycle and epidemiology :

The pathogen is seed borne (39) which is the main source of primary inoculum in a new area. Further spread is through spattering rains. A severe outbreak of the disease was observed in Maharashtra on greengram after 15 days of sowing when the crop was in 2 trifoliate leaf stage (43). It clearly indicates the seed borne nature of the pathogen. Prevalence of intermitant rains and cloudy weather during second half of July are reported to favour the disease spread. In India, the relative distribution and the severity of the disease needs to be worked out so as to understand the role of environment or other factors in its absence in kharif crops in north India, and why the disease has not assumed serious proportions as in Ohio, U.S.A.

Host resistance :

patel and Jindal (32) screened 23 varieties of greengram under artificially inoculated conditions including cvs. T 44, Hyb 45, Pusa Baisakhi, Kopergaon, Jalgaon 781 etc. but all were susceptible to *P. phaseolicola.* However, last three varieties have been rported to be resistant or moderately resistent in Maharashtra state (43).

Control :

Not much work has been done on this aspect and use of disease free seed seems to be the good alternative method. Seed treatment with hot water, mercuric chloride or streptomycin may eradicate the pathogen. Crop rotation and alteration of sowing date so that the least favourable weather conditions for the pathogen prevail, are effective in reducing the disease.

3.3 BIOTECHNIQUES :

Tissue culture : Singh and Thind (50) suggested use of somatic cell culture technique as a supplement tool for screening greengram varieties resistant to *Xanthomonas campestris* pv *vignae-radiatae*. Callus cultures were established from hypocotyl segments excised from exenic seedlings of 5 greengram varieties on MS medium fortified with 2, 4-D (2 mg/l) and Kin (0.2 mg/l). Callus was subcultured on the same medium having different concentrations of culture filtrate of the pathogen for 60 days. Culture filtrate up to 20% inhibited further proliferation while 10% was lethal. Preliminary experiment on *in vivo* screening of seedlings and on cell suspension cultures also showed similar results.

Agrobacterium tumefaciens : Multiplication of the DNA of *A. tumefaciens* has been tried in seedlings of greengram. It did not replicate under DNA filter hybridization and DNA solution enrichment reassociation techniques. These results fail to agree with the reports that large pieces of DNA of the plant genome of bacterial DNA treated plants is made up of bacterial donor of DNA (21).

It has been observed that the tumer initiation enhancer isolated rom the extract fractions of sunflower plants susceptible to *A. tumefaciens* strongly stimulates initiation of adventitious roots in the ntact plants of greengram. In cuttings also, roots were initiated when the active compound was applied after IBA treatment (7).

REFERENCES

1. Abbaiah, K., Subba Rao M., Seenaiah, P. and Satyanarayana A. 1988. Evaluation of mungbean genotypes against bacterial leaf spots. *Indian Journal of Pulses Research* 1 : 67-68.

2. Abbaiah, K. 1989. Epidemiology of bacterial leaf spot of mungbean. *National Symposium New Frontiers on Pulses Research and Development, Kanpur* 116 p.

3. Agrawal, P.C., Metha S.L., Patel P.N. and Jindal J.K. 1981, Variability in Xanthomonads of grain lequmes: Comparison of soluble protein patterns by disc gel electrophoresis. *Indian Phytopathology* 34 : 442-448.

4. Anonymous, 1960. Work of I.R.C.T. 1957-1960. *Cot. et Fibr. Trop.* 14 (2) : 83-321.

5. Behniwal, M.S. and Parashar R.D. 1979. Studies on bacterial leaf spot of mung (*Vigna radiata L.). Geobios* 6 :26-27.

6. Behniwal, M.S. and Parashar R.D. 1981, Evaluation of antibiotics and fungicides for the control of bacterial leaf spot of mung. *Haryana Agricultural University Journal of Research* 11 : 36-39.

7. Bouckaert-Urban, A.M. and Vendrig J.C. 1982. Influence of crown gall initiation enhancer on adventitious rooting. *Zeitschrift fur pflanzenphysiologie* 108 : 247-254.

8. Burkholder, W.H. 1924. Varietal susceptibility among beans to the bacterial blight. *Phytopathology* 14 : 1-7.

9. Chand, J.N., Yadav O.P. and Yadav H.C. 1977. Reaction of genetic stock of mungbean to bacterial blight (*Xanthomonas phaseoli*). *Indian Phytopathology* 30 : 565.

10. Chand, J.N. 1986. Bacterial diseases of legumes and their management. *Indian Journal of Mycology and Plant Pathology* 16 : 1-29.

11. Dye, D.W., Bradbury J.F., Goto M., Hayward A.C., Lelliott R.A. and Schroth M.N. 1980. International standards of naming pathovars of phytopathogenic bacteria and a list of pathovar names of pathotype strains. *Review of Plant Pathology* 59 : 153-168.

12. Fang, C.T., Chen H.Y. and Chu C.L. 1964. A comparative study of the species of the genus *Xanthomonas* from leuminous plants. *Acta Phytopatholggica Sinica* 7 : 21-31.

13. Gaur, R.B. and Ahmed S.R. 1983. Studies on the chemical control of bacterial blight of mung incited by *Xanthomonas phaseoli* (Smith) Dowson. *Pesticides* (17 (2) : 22-23.

14. Gaur, R.B., Ahmed S.R. amd Kataria P.K. 1984. Controlling *Xanthomonas phaseoli* (Smith) Dowson of moong seed through chemotherapy and heat therapy. *Hindustan Antibiotics Bulletin* 26 : 23-26.

15. Gurha, S.N. and Misra D.P. 1982. Donors for resistance in blackgram (Vigna mungo (L.) Hepper) against *Xanthomonas phaseoli* (Smith) Dowson. *Madras Agricultural Journal* 69 : 139-140.

16. Jindal, J.K. and Patel P.N. 1981. Variability in Xanthomonads of grain legumes, III. Variation in sensitivity to bacteriophages, *Phytopathologische Zeitschrift* 100 : 97 - 100

17. Jindal, J.K., Patel P.N. and Khan A.M. 1981. Variability in Xanthomonads of grain legumes. II. Pathogenic variability in *Xanthomonas phaseoli* mungbean strain, *Xanthomonas vignicola* and *Xanthomonas phaseoli* var. *sojense, Phytopathologische Zeitschrift* 100 : 1-9.

18. Jindal. J.K. and Patel P.N. 1984. Variability in Xanthomonads of grain legumes. IV. Variation in bacteriological properties of 83 isolates and pathogenic behaviour of cultural variants. *Phytopathologische Zeitschrift* 110 : 63-68.

19. Jindal, K.K. and Thind B.S. 1987. Control of ***Cercospora*** and bacterial leaf spots of greengram. *Indian Journal of Agricultural Science* 57 : 372-375.

20. Jindal, J.K. and Patel P.N. 1988. Bacterial diseases of pulse crops. In "Pulse Crops" (B. Baldev, S. Ramanujam and H.K. Jain eds.). Oxford and IBH Publishing Co. Pvt. Ltd., New Delhi, 316-337 pp.

21. Kado, C.I. and Lurquin P.F. 1975. Studies on *Agrobacterium tumefaciens.* IV. Non replication of the bacterial DNA in mungbean *(Phaseolus aureus). Biochemical and Biophysical Research Communications* 6[illegible] : 175-183.

22. Kore, S.S. and Shirshikar S.P. 1980. Investigations on bacterial leaf spot disease of urdbean (*Phaseolus mungo*). *Indian Journal of Mycology and Plant Pathology* 10 : 142-145.

23. Marimuthu. T. and Kandaswamy T.K. 1980. Bacterial leaf blight of greengram (*Vigna radiata var. aureus*) caused by *Xanthomonas phaseoli* and reaction of certain varieties to the disease, *Madras Agricultural Journal* 67 : 413.

24. Marimuthu, T. and Kandaswamy T.K. 1980. Permeability alteration in moderately resistant and susceptible variety of greengram infected by *Xanthomonas phaseoli. Current Science* 49 : 667-668.

25. Marimuthu. T. and Kandaswamy T.K. 1981. Toxin production *in vitro* and *in vivo by Xanthomonas phaseoli* (Smith) Dowson, the incitant of bacterial leaf blight of *Vigna radiata* var. *aureus* (Roxb.). *Madras Agricultural Journal* 68 : 121-123.

26. Marimuthu, T. and Kandaswamy T.K. 1981. Role of soluble sugars in the resistance and susceptibility of greengram varieties to bacterial leaf blight caused by *Xanthomonas phaseoli. Current Science* 50 : 643-645.

27. Marimuthu, T. 1983. Effect of sulphur containing amino acids on growth, toxin production of *Xanthomonas pamestris pv. phaseoli* and leaf blight development in greengram. *Indian phytopathology* 36 : 47-51.

28. Marimuthu, T. and Kandaswamy T.K. 1983. Changes in the amino nitrogen and amino acids in the mung lines moderately resistant and susceptible to bacterial leaf blight organism. *Indian Phytopathology* 36: 345-348.

29. Marimuthu, T. 1984. Changes in oxidative enzymes of resistant and susceptible varieties of greengram *(Vigna radiata (L.)* Wilczek) inoculated with *Xanthomonas phaseoli, Agricultural Research Journal of kerala* 22 : 100-102.

30. Patel, P.N. and Jindal J.K. 1970. Bacterial diseases in seed legumes in 1968 and 1969. Paper presented at the fourth Annual Workshop Conference of Pulse Crops at Ludhiana, April 1970.

31. Patel, P.N., Jindal J.K. and Shekhawat G.S. 1971. Study on bacterial diseases of legumes in India. Second *International Symposium on Plant Pathology, IPS New Delhi*, 149 p.

32. Patel P.N. and Jindal J.K. 1972. Bacterial leaf spot and halo blight disease of mungbean *(Phaseolus aureus)* and other legumes in India. *Indian Phytopathology* 25 : 517-525.

33. Patel, P.N., Jindal J.K. and Singh D. 1972. Studies on resistance in crops to bacterial diseases in India. IV. Resistance in mungbean *(Phaseolus aureus)* to *Xanthomonas phaseoli. Indian Phytopathology* 25 : 526-529.

34. Rangaswamy, G. and Gowda S. 1963. Some bacterial diseases of ornamentals and vegetables in Madras State. *Indian phytopathology* 16:74-85.

35. Rani, N. and Parashar, R.D. 1981. Studies on nitrogen metabolism in moong as affected by bacterial leaf spot. *Haryana agricultural University Journal of Research* 11.

36. Sabet, K.A. Ishag. F. and Khalil. O. 1969. Studies on the bacterial diseases of Sudan crops. VII. New records. *Annals of Applied Biology* 63 : 357-369.

37. Saini, L.C. and Parashar R.D. 1982. Suitability of stable bleaching powder and other chemical formulations for the control of bacterial leaf spot of greengram. *Indian Journal of Agricultural Sciences* 52 : 330-334.

38. Sandhu, D.S. 1977. Breeding for yellow mosaic resistance in mungbean. *First International Mungbean Symposium, AVRDC, Taiwan*, 176-179 pp.

39. Schmitthenner, A.F., Hoitink H.A.J. and Kroetz M.E. 1971. Halo blight of mungbean incited by a new strain of *Pseudomonas phaseolicola*. *Phytopathology* 61 : 909.

40. Schroth, M.N., Vilma Vitanza B. and Hildebrand D.C. 1971. Pathogenic and nutritional variation in the halo blight group of fluorescent Pseudomonads of bean. *Phytopathology* 61 : 852-857.

41. Sharma, H.C., Khare M.N., Joshi L.K. and Kumar S.M. 1971. Efficacy of fungicides in the control of diseases of kharif pulses-mung and urd. Paper presented in Fifth All India Coordinated Workshop on Kharif Pulses, 2 pp.

42. Shekhawat, G.S. and Patel P.N. 1977. Seed transmission and spread of bacterial blight of cowpea and leaf spot of greengram in summer and monsoon seasons, *Plant Disease Reporter* 61 : 390-392.

43. Shirsat, A.M., Kale V.V. and Pawar, N.B. 1976. Incidence of halo blight caused by *Pseudomonas phaseolicola* (Burk.) Dowson on greengram *(Phaseolus aureus)* in Maharashtra state. *Journal of Maharashtra Agricultural University* 1 : 314.

44. Singh, D. and Patel P.N. 1977. Studies on resistance of crops to bacterial diseases in India. VIII. Investigations on inheritance of resistance to bacterial leaf spot and yellow mosaic diseases and linkage if any with other characters in mungbean. *Indian Phytopathology* 30 : 202-206.

45. Singh, Gurdip, Kapoor Shashi and Singh K. 1987. Multiple disease resistance in mungbean with special emphasis on yellow mosaic virus. *Second International Symposium on Mungbean, Bangkok, Thailand* 25-26 pp.

46. Singh, P.P. and Thind B.S. 1981. Histological changes and enzyme production in mungbean in response to infection with *Xanthomonas phaseoli f. sp. Vignaeradiatae* and *Cercospore canescens. Third International Symposium on Plant Pathology, IPS. New Delhi.* 202-203 pp.

47. Singh, P.P. and Thind B.S. 1984. Interaction between *Xanthomonas campestris* pv. *vignaeradiatae* (Sabet *et al.)* Dye and *Cercospora canescens* Ell. et Mart. on mungbean in relation to disease development. *Phytopathologia Mediterranea* 23 : 75-76.

48. Singh, P.P. and Thind B.S. 1978. Histological changes in greengram leaves in response to *Xanthomonas campestris pv. vignaeradiatae and Cercospora canescens* infection. *Indian Phytopathology* 40 : 149-151.

49. Singh, Sukhwinder and Thind B.S. 1988. Biocontrol of *Xanthomonas campestris pv. vignaeradiatae* infection on greengram. *Indian Journal of Mycology and Plant Pathology* 18 : 111.

50. Singh, Sukhwinder and Thind B.S. 1988. *In vitro* selection of cells resistant to *Xanthomonas campestris pv. vignae-radiatae* in mungbean. *Indian Journal of Mycology and Plant Pathology* 18 : 108.

51. Singh, Sukhwinder and Thind, B.S. 1989. Factors affecting the antagonistic activity of the microorganisms against *Xanthomonas campestris pv. vignaeradiatae in vitro, Indian Phytopathology* 42 : 280-281.

52. Soni. P.S. and Thind, B.S. 1987. Detection of *Xanthomonas campestris pv. vignicola* (Burk.) Dye from cowpea seeds and *Xanthomonas campestris pv. vignaeradiatae* (Sabet et al.) Dye from mungbean seeds. *Phytopathologia Mediterranea* 26 : 1-6.

53. Sood, A.K., Thind B.S. and Singh N. 1976. Occurrence and control of bacterial leaf spot of mung (*Phaseolus aureus* Roxb.) in the Punjab state. *Journal of Research Punjab Agricultural University* 13 : 182-185.

54. Sood, A.K., Thind, B.S. and Singh, N. 1976. *In vitro* evaluation of some antibiotics and fungicides against *Xanthomonas phaseoli,* causal agent of leaf spot of mung *(Phaseolus aureus). Hindustan Antibiotics Bulletin* 18 : 119-120.

55. Tanii, A. and Baba T. 1979. Bacterial stem rot of adzuki bean (*Phaseolus radiatus var. aurea Prain)* caused by *Pseudomonas adzukicola A. Tanii & T. Baba Nov. Sp. Bulletin of Hokkaido Prefectural Agricultural Experiment Station No.* 42 : 29-42.

56. Thakur, R.P., Patel P.N. an Verma J.P. 1977. Studies on resistance in crops to bacterial diseases in India. *Indian Phytopathology* 30 : 217-221.

57. Thakur, R.P. Patel P.N. and Verma J.P. 1977. Genetical relationships between reactions to bacterial leaf spots, yellow mosaic and cercospora leaf spot diseases in mungbean *(Vigna radiata). Euphytica* 26 : 765-774.

58. Thakur, R.P., Kumar S., Patel P.N. and Verma J.P. 1978. Studies on virulence of biochemical mutants of *Xanthomonas phaseoli mungbean strain. Indian Phytopathology* 31 : 52-56.

59. Thind. B.S., Kumar K. and Soni P.S. 1984. Eradication of *Xanthomonas campestris pv. vignaeradiatae* from mungbean seeds by physical and chemical treatments. *Indian Journal of Nycology and Plant Pathology* 14 : 81.

60. Thind B.S. and Kumar K. 1985. Chemical Control of bacterial leaf spot and cercopora leaf spot of munglian. Indian phytopathology 38:601.

61. Verma, M.L. 1985. Control of foliar diseases of urdbean *Vigna radiata var. mungo). Indian Phytopathology* 38 : 600-601.

62. Yadav, H.C., Chand J.N. and Saharan G.S. 1981. Inheritance of resistance in mungbean to bacterial leaf spot. *Fifth International Conference on Plant Pathogenic Bacteria Hissar* 1981, 580-583 pp.

63. Zaumeyer, W.J. and Thomas H.R. 1957. A monographic studies of bean diseases and methods for their control. *U.S.D.A. Technical Bulletin No.* 868.

CHAPTER 4

VIRUS AND MYCOPLASMA DISEASES

Large number of viruses infect the mungbean and urdbean crops but lot of confusion exists in the literature owing to insufficient information or incomplete studies made. In most cases, new viruses are described and named on the basis of symptoms only which is not adequate. In several cases, the identification implicating known described viruses has often been based on symptoms alone. Though, the symptoms are very important, they are notorious inadequate guide to identification. Different strains of the same virus may induce widely different symptoms in the same cultivar; a virus may cause different symptoms in different cultivars and unrelated viruses may cause similar symptoms in a common host. For example, yellow mosaic symptoms are produced by atleast 4-5 viruses but they are not related with each other necessarily. Large number of unrelated viruses produce mosaic and mottling symptoms in mungbean and urdbean. Identification therefore requires great caution and should be confirmed after studying maximum properties particularly the particle morphology, transmission, biophysical properties, serology, host range and the reaction of diagnostic species to infection, purification and the properties of particles which require sufficient laboratory facilities and technical know how. Important virus diseases of greengram and blackgram have been reviewed by Raychaudhury (197), Chenulu and Varma (55), Amin and Singh (14), Sen Gupta (206) and PANS (173).

4.1. VIRUSES CAUSING YELLOW MOSAIC AND ALIKE SYMPTOMS :

Yellow mosaic and alike symptoms are caused by at least 5 viruses which are different from each other in one or more of their characters like properties, host range, transmission, serological reaction etc.

4.1.1. MUNGBEAN YELLOW MOSAIC (INDIAN STRAIN) :

Yellow mosaic of mungbean was observed occurring in India as early as 1945 (17, 289). The disease was later studied in detail by Nariani (156) who named it `yellow mosaic' of *Phaseolus aureus* and reported it to be different from the yellow mosaic of *Phaseolus lunatus* and *Dolichos lablab* described previously (48). Yellow mosaic on blackgram was also noted under natural conditions in India (298). The disease occured seriously in the Tarai area of Uttar Pradesh state of India where Nene (159, 161) studied it in detail and named the cause as `Mungbean Yellow Mosaic Virus' (MYMV) . The disease is very much devastating and is reported to cause serious losses to mungbean as well as urdbean in several countries of the world particularly India, Pakistan (9) Bangladesh (11), Ceylon (168, 218), Philippines (26), Indonesia (99), Thailand (278), Papua New Guinea (175), Columbia, Ethiopia and in South-East Asian Countries (178). In India, the disease is prevalent in states of Uttar Pradesh, Madhya Pradesh, Bihar, Punjab, Haryana, Himachal Pradesh, Rajasthan, Orissa, Andhra Pradesh and Tamil Nadu (15). The detailed studies indicate that the mungbean yellow mosaic viruses reported from Philippines and Thailand are different from the one occurring in India, Pakistan and several other countries.

Nature and extent of damage :

Out of all the diseases occurring on greengram and blackgram, the yellow mosaic is most destructive and wide spread. It is reported to cause much damage to mungbean crop as compared to urdbean in Haryana possibly due to choice of varieties grown in the area (239).

The disease causes significant reduction in dry weight of plant, length of shoot and root, the leaf area and pod bearing capacity of plants of both the crops (238). The reduction in plant height and shoot weight may be to an extent of 38.2 and 28.5 % respectively (50). The chlorosis, stunting and reduced branching due to the disease, also, contribute to the losses in yield which may range from 10 to 100% (15, 50) depending on the host variety and the age of plant at the time of inoculation. For example, reduction in pods/plant, pod weight, 100 grain weight and yield due to MYMV infection is more in mungbean variety PIMS 3 as compared to PIMS 2 or PIMS 1 (46). Age of the plant at the time of inoculation is an important factor controlling the yield reduction caused by MYMV (9). The urdbean plants infected up to 3 weeks after sowing, do not produce any seed but the losses may be only negligible if plants are infected after 8 weeks of sowing (146). Losses may range between 15-85% in plants infected from 4 to 8 weeks after sowing. The main yield component affected by the disease is pod/plant (146) but seeds/pod and the seed weight are also considered as the important components affected by the disease (21, 297).

Not only yield but the quality of seed produce is also affected adversely due to the disease. The colour, texture and size of seed is deteriorated but the germination capacity is not much affected (50, 297). However, germination of the affected seeds may be delayed to some extent. The reduction in seed weight may be up to 25.7% (50). The nutritive value of seed is also influenced by the disease. Infection of urdbean plants with in 10-30 days after sowing causes increase in total amino acids and protein content of seed but it does not happens if the plants are infected after 70 days of sowing (36). Of 15 amino acids detected, 9 are reported to increase, 4 are decreased while 2 remain unchanged.

Symptoms :

In mungbean, the first symptoms appear on young leaves in the form of mild scattered yellow specks or spots after 12-16 days of inoculation (156). The next trifoliate leaf emerging from the growing apex shows alternating yellow and green patches (Fig. 4.1). The leaf

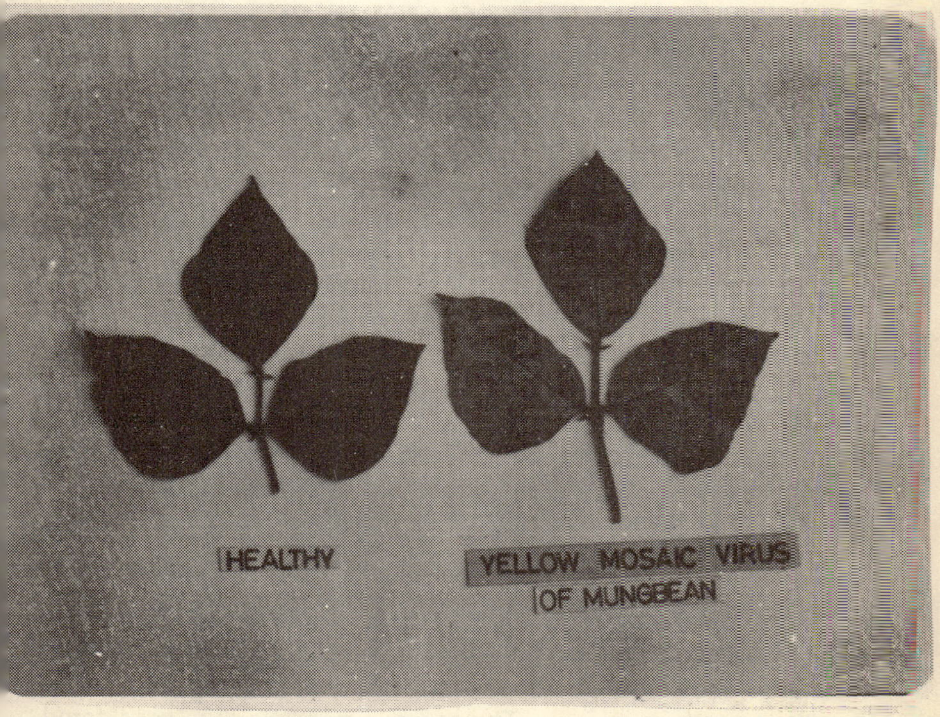

Fig.4.1
Yellow mosaic of greengram.

size is not affected much but some times the green areas on the affected leaf may be slightly raised resulting in slight puckering and reduction in the leaf size. In the newly produced leaves, the yellow areas increase in size ultimately turning some of the apical leaves completely yellow.The disased plants are stunted with very few flowers and pods and mature late. The pods are curled and small with immature,shrivelled and small seeds. The proportion of the affected seeds increases with the increasing severity of the disease. In some of the varieties like Jalgaon 781, Kopergaon and T 51, the seeds may develop mottling with yellowish tint. The early infection of plants may cause more severe symptoms as compared to late infectons (228). It has been noted in Pakistan, that the bright yellow blotches often become whitish in older leaves and under severe infections only few poorly filled pods or rudimentary scales bearing no seeds are produced (9). Symptoms on blackgram are similar to those produced on greengram but on other hosts they may differ (156).

Nene (159, 161) observed two types of symptoms of yellow mosaic and described them under the names `yellow or Chlorotic mottle' and `Necrotic mottle'. The yellow mottle symptoms appear more often nearer to flowering period in the form of yellow, some what diffused spots which are mostly round and scattered on the leaf lamina. Spots increase rapidly forming yellow patches alternating with the green colour. New emerging leaves show sign of mottling right from the beginning and typical symptoms are visible as they are unfolded. Affected leaves are reduced in size which later turn yellow and then necrotic. The necrotic mottle appears in the form of small yellow spots well delimited by finer veins. Several adjoining spots may form continuous yellow zones on the affected leaf. Very soon, a necrotic spot develops in the centre of yellow spot and the yellow colour gradually changes in to whitish shade. These symptoms are systemic. The disease does not cause reduction in size of the leaf and pods and the pod bearing capacity of the plant.

In fact, the yellow mottle and necrotic mottle are the two types of the symptoms produced on different varieties by MYMV (145). The varieties showing necrotic mottle reation, yield better than those with yellow mottle symptoms. The necrotic mottle is thus considered a type of resistant reaction.

Virus :

Mungbean Yellow Mosaic Virus (MYMV) Nene 1968
Synonym : Yellow Mosaic of *Phaseolus aureus* Narian 1960
Yellow Mosaic Virus (YMV) Ahmed and Harwood 1973.

Morphology : The Indian isolate of MYMV has not been studied in detail but as per personal communication from Dr. S.P.S. Beniwal (218), it contains the spherical particles with a diameter ranging from 80 to 100 nm.

Host range : In India, MYMV has been reported to infect several plant species in addition to *Vigna radiata*. They are : *Vigna mungo* (L.) Hepper, *Glycine max* (L.) Merrill, *Phaseolus acutifolius* (A.) Gray, *P. aconitifolius* Jacq, *P. lathyroides* L. and *'Dolichos biflorus* L. (156). At Pantnagar, India, several plant species were tested and *Brachiaria ramosa* (L.) Staff, *Eclipta alba* Hassk, *Xanthium strumarium* L. and *Cajanus cajan* (L.) Millsp. were found the additional hosts of MYMV (164, 166). Later, *Cosmos bipinnatus* Cav. and *Phaseolus vulgaris* var. Manitov were also infected by the virus (192). Though, Narian (156) reported *Phaseolus lunatus* not to be infected but positive results were obtained by Nagraj *et al.* (141). They further reported that the virus could infect *Phaseolus coccineus* L., P. multiflorus Wild species P-1, *P. panduratus* Mart ex Benth, *P. pilosus* HB & K., *P. atropurpureus* Moc & Sess., *P. odenanthus* G.F.W. Mey Prim. FL Esseq, *P. bracteatus* Nees & Mart., *P. angularis, P. phyllanthus* and *P. polystachynus* also. A survey has shown that at least 75 weed hosts of this virus are there in nature which may act as a reservoir (197). In Madhya Pradesh, wild mung (*Phaseolus trilobus*) has been found infected by MYMV under natural conditions (110).

Tranmission : The graft transmission of the virus was successful (9, 156). The virus is transmitted mainly by white fly, *Bemisia tabaci* Genn. and not by sap, seed, soil or nematodes (9, 143, 156). The transmission tests with some species of leaf hoppers, aphids, thrips and beetles were not successful (9).

The relationship between the MYMV and its vector, *B. tabaci* has been studied in detail. It is able to acquire the virus within a minimum period of 15 minutes but the transmission efficiency of the vector increases with the increase in acquisition feeding period reaching 90.9% in case of 4 h acquisition feeding (143, 165, 196).

Some workers have reported maximum transmission of MYMV after an acquisition feeding of 24 h (54, 138, 156).The minimum acquisition period may increase if the variety of source and test plants are different from each other (161, 196). A four h starvation prior to acquisition feeding increases the efficiency of MYMV transmission by *B. tabaci* (138, 144, 145) but negative results have also been reported on this aspect (54). A white fly adult can acquire the virus from the infected mungbean plant as early as 1-3 days before appearance of symptoms (193). A white fly fed on diseased plant extract may survive for some time but is unable to transmit the virus in blackgram plants.

The minimum incubation period of MYMV in the vector *B.tabaci* ranges from 4-5 h (144, 161), to 5-6 h (54) and 8 h (138). The transmission efficiency of the vector increases with the increase in inoculation feeding period. The minimum inoculation feeding period required for 100% transmission is reported differently as 1 h (54), 4 h (143) and 24 h (138, 156). It may vary according to the variety of source and test plants used (196). Howevewr the vector may infect 50-70% plants within 10-15 munites of inoculation feeding only.

The MYMV appears to be circulative in *B. tabaci* where it persists for a maximum of 10 days in females and 3 days in males (193). It has been confirmed by detecting the virus in first 2-3 abdominal segments of the viruliferous white fly by using fluorescent dye (acridine orange) which stains the virus as dark orange mass (122). Though, the MYMV may be retained by the white fly until its death (4-10 days) but the transmission capacity of the vector decreases with the advancing period of retention (144). The transovarian carryover of MYMV has not been observed in *B. tabaci* (9, 138, 144,193). Sex of the vector also determines the transmission efficiency of the virus. The female vectors remain alive and infective for a longer period (8-10 days) as compared to male (4-5 days). Beside it, the percentage of adults capable of transmitting the disease, the transmission capacity and survival of the viruliferous adult on host after inoculation are much higher in case of females as compared to males (193, 195). Similarly the nymphs of the vector can acquire and transmit the MYMV (193) but first instar nymphs are unable to acquire the same (138). The life history of the vector, its handling and maintenance has been studied in detail for use in virological research work (161, 194).

Host parasite interaction :

After infection by MYMV, the histopathological changes occur in the host and its physiology and biochemistry is altered to a considerable extent. The epidermal cells of the infected mungbean leaf when treated with fluorescent dye, acridine orange, the virus appears as orange-red fluorescing inclusions in the cytoplasm of cells (122). The pyronine G methyl green combination proved superior over other strains for differentiation between virus and the host nucleus.

The MYMV infection causes reduction in the chlorophyll content and the functional chloroplast cells in the leaves resulting in reduced rate of photosynthesis (21, 81).Reduction has been observed in chlorophyll `a' and `b' and carotenoid (240). The chlorophyll is reduced with increasing disease severity and it is assigned to the reduced activity of the chlorophyllase (183). The reduction in total dry matter, chlorophyll and carotenoid has been recorded in leaves of resistant as well as susceptible varieties of blackgram. The disease causes increase in the respiration rate. It decreases the starch while sugar contents are increased in the beginning with a decrease afterward (21, 81, 113). The disease results in decreased protein and DNA contents but RNA and soluble nitrogen are increased to some extent (21, 240). It also influences the amino acid contents of the leaf. In addition of six amino acids also contained in the healthy leaves, cystine, alenine and leucine are detected in the diseased leaves (200). The infection of mungbean decreases calcium in addition to sugar. It ultimately causes severe infestation of diseased plant of mungbean by a non vector leaf hopper, *Empoasca kerri* (268). The auxin content of the infected plant is reduced which reduces the leaf size with stunting of urdbean plant (37). Tripathi *et al*. (282) observed higher activity of polyphenoloxidase and peroxidase enzymes in mungbean leaves infected by MYMV. On the other hand, the disease causes significant decrease in total phenols; orthodihydric phenols; free amino acids; total, soluble and reducing sugars in mungbean leaves (133) while non reducing sugars increase slightly. Infact, physiological changes in the infected plants do not give a complete insight of the physiological derangement in the host plant (240). However, it is clear that the increased rate of the respiration, reduces photosynthesis and other physiological disorders change the biochemistry of the infected plant resulting in retardation of growth, delay in fruiting and reduced yield (81).

The virus is reported not to cause any variation in the strain of root nodule bacterium (290).

Disease cycle and epidemiology :

The MYMV is neither sap nor seed or soil transmissible but through white fly only. It has a very wide host range including more than 75 weed and crop plant species which may serve as the reservoir of the virus and are considered as source of primary inoculum. Pigeonpea stands in the field for a longer period of the year and may be one of the important source (166). Further spread of the disease is mainly through white flies which start spreading the virus with the onset of monsoon in Northern India. During a test, white flies collected before onset of the monsoon were not found viruliferous whereas those collected after the monsoon, infected nearly 75% test plants (287).

The further spread and development of the virus disease depends on the efficiency of the vector. *B. tabaci* is very efficient vector acquiring and inoculating the virus within 10-15 minutes only. It remains viruliferous for several days after acquisition. A singl viruliferous whitefly adult is able to transmit the MYMV successfull though 100% transmission takes place when 10 whiteflies per plan are used (54, 138, 165). The vector is seen in nature, all the yea round, on several host plants and is able to move long distance wit the wind current (255).

Host factors also influence the spread and development o MYMV in nature. The rate of disease spread differs from variety t variety; for example in mungbean varieties PIMS 3 and PIMS 4 th disease spreads much faster as compared to PIMS 1 and PIMS (46). Under field conditons, the spread of the disease in varietie with 'Yellow mottle' symptoms is faster than those with 'Necroti mottle' symptoms. This difference may be due to the differences i acquisition feeding period or due to the differential nutritiona compatibility of the host variety and the virus (203). Primary as we as first trifoliate leaves are suitable for infection by *B. tabaci* (54) Pre-and post-inoculation exposure of seedlings to dark and th stages of the crop at the time of inoculation do not influence the pe cent transmission of MYMV by *B. tabaci* (145).

Environment is the important factor which affects the severity of MYMV by influencing the disease development as well as the activity of the vector. The mungbean leaves showing early evidence of the infection may recover under cool growing conditions suggesting that yellow mosaic requires sustained high temperature (9). The incidence of MYMV is reported much higher in the crop sown in June-July as compared to that sown in March-April (288) in Haryana where temperature levels are higher in March-April as compared to June-July. The incubation period of MYMV in seedlings is extended and the transmission per cent is reduced during November-December when temperature is lower as compared to July and October with higher temperature levels (145). Similarly under South Indian conditions also, higher whitefly population with high incidence of MYMV has been observed on April sown crop (140). Based on these observations, a positive correlation was established between the maximum temperature and white fly population and the disease incidence. Accordingly, it is possible to predict the whitefly population and the disease incidence, one week in advance, using multiple regression analysis technique in which maximum temperature is an important factor for whitefly population while maximum temperature, rainfall and whitefly population are important in predicting the disease out break (140). Increase in the acreage of the susceptible crop varieties and unusually dry condtions may allow a population build up of the vector *B. tabaci* resulting in heavy out break of MYMV (278). On the other hand in Bangladesh, high incidence of the disease has been observed on winter crop as compared to summer crop of mungbean (11).

Host resistance :

Source of resistance : Efforts have been made to search out the sources of resistance in germplams as well as established varieties of greengram and blackgram and variable response has been observed (135, 142, 156, 161, 256 etc.). Reports on this aspect are summarized in Table 4.1.

Table 4.1

Source of resistance in greengram and blackgram against Mungbean Yellow Mosaic Virus

Host species	Entries tested	Entries showing resistance	Reference
Greengram (*Vigna radiata*)	215	None	(161)
	-	Line 5908-2, 6410, 6501, 670	(73)
	496	P 346-68, P 366-68	(72)
	7	No. 54 least susceptible	(288)
	157	None, 6 local collections tolerant	(8)
	435	15229 and L 24-2	(85)
	122	25 lines resistant	(270)
	18	LM 220, PLS 274 were free but susceptible in glass house	(294)
	620	ML 6, 15060, 15225, 15227, 1788/ 7, EC 27574, Accession 392, 404, 461	(296)
	-	L 24-1, 15227	(87)
	-	L 24-2,15225,15227, Hyb 4-3A 12-4, ML62, ML65	(14)
	504	11 lines least susceptible to MYMV and whitefly	(120)
	713	ML 109, ML 161 were free, ML 111, ML 131, have traces, LM 87, LM 113, LM 134, 170, 171, 168, 238, 392, 404, 496, 509, 687, 692	(203)
	248	Tarai local resistant and 21 lines moderately resistant	(172)
	-	Six from gamma ray irrediated lines	(209)
	-	Selection 293-1-1	(188)
	13	Pusa Baisakhi least susceptible	(54)
	4	ML 1 & ML 3 resistant to MYMV and whitefly	(57)
	62	G 65, ML, 25, S 8, T 44, T 192-1, ML 1, 7, 17, 24, 25, 35	(121)
	7	ML 1, ML 3, ML 5 and ML 270 resistant to MYMV and white fly	(56)
	-	ML 26, ML 9, 11157, ML 70-10, L 24-2, LM 896, 15225 & 15229	(101)
	157	11/99,11/395, 16-303-20-3-g	(234)
	-	Hyb 4-3A & Hyb 12-4 were immune	(235)
	198	P Nos. 20, 68, 82, 84, 86, 94, 332, 367, 369, 453, 670, EC 103127, H 70-6, LM 150, LM 172,LM 290. NO. 525 and China selection	(52)
	4380	ML lines 234, 268, 269, 272, 274, 338, 360, 369, 371; LM Nos. 113, 687, 695, 696, 744; MUG Nos. 125, 130, 132 to 136, 138, 139, 145, 170, 173, Tarai local, AVRDC lines V 2182, 2294, ML 266, ML 267,337, 375, Hyb 4-3A were resistant for 6-7 years	(237)

(Contd....)

Blackgram (*Vigna mungo*)	67	Mash 1-1, Msh 41-13, No. 1-1, T 44 and D 6-7 selection	(161)
	122	25 lines	(8)
	-	Lines 128, 129, 133, 151, 296 from U.S.A. were highly resistant	(9)
	6	Mash 1-1	(288)
	281	BR 10, K 66-110, NP 6, 7, 14, 16,19, 20, 21, PLU 277, 340, U 9, 1517 and 7368/4-B were free	(90)
	-	LU 220, LU 237	(68)
	12	Mash 48, Selection 1, L 35-5	(100)
	846	UG Nos. 135, 253, 254, 255, 261, 267, 291, 292, Mash 1-1 are highly resistant. UG Nos. 117, 119, 152, 153, 157, 170, 188, 191, 192, 193, 195, 204, 203, 234, 240, 242, 247, 248, 250, 251, 260, 262, 268, 273, 283, 289, 290, 300, 355, PLU 188A, PLU 177, PLU 476, Mash 48 and Saradha Mash resistant to moderately resistant	(236)
	29	25 including Barabanki local, COBG 10, DU 1, K 66-188, NP 16 and NP 6	(174)
	177	BR 10, K 66-110, NP 6, 7, 14, 16, 19, 20, 21, PLU 63, 277, 340, 476, 557, 574, 577, U 9, 13/2, 14/5, 15/7, 31/4, 7368/4B	(257)
	975	48 entries were free	(265)
Phaseolus spp.	-	*P. calcaratus, P. trilobus*	(142)
	-	*P. sublobatus*	(188)
	-	*P. sublobatus* Pantnagar strain A	(230)

Inheritance of resistance : On the basis of analysis of F 1 and F 2 data involving mungbean variety L 24-2 resistant to MYMV, it is indicated that the resistance is controlled by a single recessive gene which is inherited independently to seed colour and maturity (231). On the other hand, the crosses involving variety 15229 showed monogenic dominant control of resistance while in case of variety L 24-2, it was controlled by 2 complementary recessive genes (204). Shukla *et al.* (220) used Tarai local (resistant) and Jawahar 45 & G 65 (susceptible) mung been parents in crosses and found resistance to MYMV to be digenic and recessive. The genotypes assigned to these varieties belonging to resistant, moderately resistant, tolerant and susceptible categories are : $s_1s_1\ s_2s_2$ ii, $s_1s_1\ S_2S_2$ II, $S_1S_1\ s_2s_2$ II and $S_1S_1\ S_2S_2$ ii respectively (219). Here $s_1\ s_2$ are duplicate recessive resistance genes and I is an inhibitor which counteracts the effect of some combinations of the dominant alleles S_1S_2. Singh (233) also found the MYMV resistance to be digenic, recessive and free from cytoplasmic effects. In case of blackgram, inheritance of rsistance to MYMV is described differentlyin different varieties. In varieties Pant U 19 and Pant U 26 it is reported to be controlled monogenitically but in both the cases different genes are there (109). In cvs. LU 220 and LU 237 the resistance to MYMV is due to a single dominant gene (68). On the other hand in cvs. Pant U 84 and UPU 2 the resistance to MYMV is controlled by two independent recessive genes which are designated as ymv-1 and ymv-2 (78, 293). Attempts have been made to transfer the resistance to MYMV from *Phaseolus sublobatus* to *Vigna radiata* which indicate that the resistance to MYMV is dominant character though number of genes responsible for resistance are not established (12).

Breeding for MYMV resistance : Breeding of resistant varieties against MYMV is necessary due to very wide host range of the virus and presence of the host plants throughout the year. The high mutation rate in the viruses makes vertical resistance less dependable and hence, horizontal resistance is favoured in such cases (203). Resistance detected in germplasm has been used to improve the varieties of both the crops. Resistance to MYMV could be incorporated in susceptible high yielding blackgram varieties with otherwise desirable characters (9). Greengram varieties resistant to MYMV with good yield and other agronomical characters have been developed and released for general cultivation (51, 97, 132, 203). Besides MYMV, resistance to its vector, whitefly, has been identified in several greengram varieties (56, 57). Varieties have been identified possessing multiple disease resistance particularly to MYMV, bacterial blight, Cercospora leaf spot (203); MYMV, Cercospora leaf spot, *R. solani, X. phaseoli* and leaf crinkle virus (101); MYMV, bacterial leaf spot, Cercospora leaf spot (237). The interspecific crosses between *Phaseolus calcaratus* and *Vigna radiata* were not successful (141) but the cross was successfully made between *Vigna radiata* and *Phaseolus sublobatus* for incorporating resistance to MYMV (12) Gill *et al* .(84) got high yielding and MYMV resistant lines from the cross between *V. mungo* and *V. radiata*.

Infact,the breeding strategies for future should be more exhaustive. Rigid screening for more reliable sources of resistance at intra and inter -species levels, use of mutational approach, transfer of resistance to develop high yielding suitable varieties following suitable breeding procedures based on genetics of resistance are important. Irradiation treatment may help in developing MYMV resistant mutants of mungbean (209).

Selection for MYMV resistance : Since the viruses are obligate in nature, the necrotic reaction produced by them may result in restriction of further development of the disease. The varieties showing necrotic mottle reaction are damaged less by MYMV as compared to those with yellow mottle. The rate of spread of the disease is also very less in varieties with necrotic mottle reaction and hence, it is considered as the resistant reaction (46, 147, 161). It has also been mentioned that resistant varieties have more number of 3-4 septate hooked hairs and thicker cuticle than the susceptible varieties which posses less number of straight, non septate leaf hairs and thinner cuticle (49). These characters of the leaf probably influence the infestation and feeding by whiteflies on the host. For scoring the MYMV severity on a variety, several scales have been used. In most cases percentage of infected plants has been used as the basis of scoring. Infact, the major setback to pod maturity and reduction in yield occurs at only 25% intensity in blackgram. Hence, it is suggested to select material for breeding programme with minimum disease (90). The scale used by Chaudhury *et al.* (52) seems to be more suitable in this context which is as follows :

Grades	*Disease severity*
Free (F)	No infection
Resistant (R)	0.1-2% infection
Moderately resistant (MR)	2.1-5% infection
Moderately susceptible (MS)	5.1-10% infection
Susceptible (S)	10.1-20% infection
Highly susceptible (HS)	Above 20% infection

However, this scale says nothing about the reaction of plants against the pathogen which is important in development of a disease like yellow mosaic. Singh *et al.* (265) used a three category scale in which time of disease appearance and the type of symptoms have been considered :

Resistant (R) = No visible symptoms

Moderately susceptible (MS) = Yellow mosaic symptoms appearing at a later stage of crop growth, less stunting.

Highly susceptible (HS) = Severe mosaic symptoms at early stage of crop growth with stunting.

For scoring the mungbean varieties in International nursary, a 5 point scale is suggested as follows :

0 = No plant infected
1 = Up to 5% plants infected
2 = 6-15% plants infected
3 = 16-25% plants infected
4 = 26-40% plants infected
5 = 41% or more plants infected

The scale recommonded by Mayee and Datar (1986) as referred under the chapter `Fungal diseases' is based on per cent of diseased plants, symptom severity and effect on plants as given below :

0 No plants showing any symptoms.
1 Less than 1% plants showing symptoms.
3 1-10% plants showing mottling of leaves.
5 11-20% plants showing mottling and yellow discolouration of leaves.
7 21-50% plants showing mottling and yellow discolouration of leaves. Stunting of plants.
9 51% or more plants affected, stunting of plants pronounced, flower and fruit set reduced. Yellow mottling severe.

At growth stages : Seedling, flowering, pod formation and pod maturity.

Control :

It is not possible to control the yellow mosaic disease by adopting a single control measure. However, use of resistant varieties is the best method which should be supported by control of vector by using chemicals or adopting cultural practices. Since, the virus causes more severe symptoms and yield losses in early infected plants, it is suggested to avoid the early infections by cultural or vector control (228).

Chemical control : The insecticide should be able to kill the vector within 10-15 minutes so as to avoid acquisition or inoculation of MYMV by *B.tabaci* (162, 164) but none was found effective. However, spray of mineral oil killed the vector within 1-3 minutes (165). Hence, spray of a mixture of power oil and emulsifiable orchard oil (2%) + endosulfan (0.025%) + oxydemeton-methyl (0.05%) has been recommended (164). Four sprays are to be done at an interval of 10 days starting from 20th day of sowing. The incidence of MYMV is reduced by weekly spray of metasystox starting from 15 day old crop (129); three sprays of anthio at an interval of 15 days starting after 15 days of germination(54); two sprays of nuvan, monocrotophos phosphamidon @ 500, 875 & 250 ml/ha respectively at an interval of 15 days(1); three sprays of monocrotophos @ 0.25 kg a.i./ha at 10 days interval starting from 15th day of sowing (139) or spray of endosulfan (0.1%) – metasystox (0.1%) but singificant increase in yield has not been achieved (194).

Soil application of solvirex at sowing time followed by 3 weekly sprays of metasystox starting after 20 days of sowing have been found quite effective (242). Application of temic or disyston in soil reduces the disease effectively (91). Soil application of granular insecticide like aldicarb or disyston @ 1-2 kg a.i./ha or with the fertilizer reduces the disease and increases the yield (88, 119, 256, 300).

Spray of insecticides with other chemicals has also been tried and found effective against the MYMV disease. For example, spray of aureofungin (0.003%) and dimecron (0.25 kg a.i./ha) controls the disease and increases the yield (10). Similarly, application of aldicarb with endosulfan and captan has been reported to reduce the disease by 31-39% (300). Extracts of different plants have also been tried. Spray of partially clarified aqueous leaf extract of *Clerodendron fragrans* (4%) at 3-4 days interval from the seedling stage is reported to reduce the disease and increase the yield while extrat of *Aerva sanguinolenta* and root extract of *Boerhaavia diffusa* delay the appearance of the disease symptoms (292).

Cultural : Manipulation in dates of sowing may help in minimizing the whitefly population and thus reduce disease incidence but it seems to differ much from region to region. In Uttar Pradesh, the incidence of MYMV on mungbean is reported to be 39.3, 20.3 and 5.0 % in sowings of 15th June, 1st July and 15th July respectively (129). On the other hand, in Tarai region of Uttar Pradesh early sown crop of mungbean had singificantly less yellow mosaic infection as compared to the late sown one (194). In Tamil Nadu state, low population of whitefly and the YMV incidence was observed on mungbean crop sown in July to September or December but the incidence was high in sowings of March to May (137). High disease in March to May sowings has been attributed to the high temperature which is favourable to whitefly. In Madhya Pradesh, the effect of sowing dates on the incidence of MYMV on mungbean from March to August was non significant (111).

The incidence of MYMV is reduced in plots with intervarietal intercropping of mungbean and urdbean but rate of the spread of the disease was much higher in intercropped plots as compared to pure stand (71). The yellow mosaic of soybean is reduced greatly by growing the barrier crop of maize or *bajara* at the rate of 4 rows of barrier crop for every 6 rows of soybean (53).

Biological control : A fungus *Paecilomyces farinosus* (Dick. ex Fr.) Brown & smith has been found parasitic on the vector of MYMV, *Bemisia tabaci*, but the practical utility of the fungus to control the whitefly is yet to be explored (163).

4.1.2. MUNGBEAN YELLOW MOSAIC VIRUS (Thailand isolate):

In Thailand, heavy out break of mungbean yellow mosaic disease was observed in 1977 (278) and the severe disease incidence was attributed to build up population of its vector, *Bemisia tabaci*, due to increased area of mungbean and prevalence of unusually dry conditions, favourable for the vector. The virus is different from the Indian isolate of MYMV in regard of particle size, transmission and the host range characters (95).

Symptoms :

Symptoms are different slightly from the Indian Yellow mosaic disease. No symptoms appear on the inoculated primary leaves. In the systemically infected leaves of mungbean, urdbean, adzuki bean and soybean, vein yellowing along the veinlets appear at the early stage and then yellow mosaic symptoms develop. First emerging trifoliate leaves of mungbean show severe down ward curling (95).

Virus :

The virus like particles (VLP) have been found associated with the yellow mosaic disease of mungbean in Thailand. The VLP are isometric, 15 -20 nm in diameter forming loose aggregates in the nucleus of the infected phloem cells (277). In vacuoles or lumena of the partially or fully differentiated sieve elements, VLPs occasionally form aggregates having a double cylindrical arrangement of particles. The purified virus has geminate particles of 18-30 nm size. The nucleic acid isolated is identified as circular, single stranded DNA with a molecular weight of 8×10^5. The thermal inactivation point of the virus is 40-50°C for 10 minutes ; dilution end point 10^{-2} to 10^{-3} and longevity *in vitro,* 1-2 days at 20°C. The purified virus preparations had a UV light absorption spectrum typical of a nucleoprotein with a A_{260}/A_{280} value of 1.3-1.4 (93, 95).

Transmission : In the beginning, the virus was not transmitted mechanically but later it could be transmitted by conventional inoculation techniques (95). The highest transmission rate being obtained with 0.1 M potassium or sodium phosphate buffer of pH 7.8. The virus was transmitted by whitefly, *Bemisia tabaci* from mungbean to mungbean, soybean and top crop bean. Generally, the whitefly transmitted viruses are not transmitted mechanically but this seems to be the exception. Indian isolate of MYMV is not transmitted mechanically.

Host range : Out of 26 plant species tested, only 7 species belonging to Leguminosae were infected showing mainly yellow mosaic or leaf curling. It has not been observed in nature on blackgram. The host range of the Indian isolate is broad as it infects *Brachiaria ramosa* and *Eclipta alba* etc. while Thailand isolate is restricted to Leguminosae only.

Host pathogen interaction :

The virus like particles form loose aggregates some times filling the total nuclear volume of the infected phloem cell (277). These nuclei are hypertrophied in which aggregates of VLP and the fibrilar bodies are observed as early as 2 days before symptom appearance. In vacuoles or lumena of the partially or fully differentiated sieve elements, the particles occasionally form aggregates having a double cylindrical arrangement. No particles are detected in tissues other than phloem of infected plant.

Disease cycle and epidemiolgy :

No detailed work has been done except that the disease is transmitted mechanically as well as by whiteflies. The optimum temperature for the symptom expression of the disease has been shown to be 25-30°C in growth chamber or 30°C day/20°C night in the green house (93, 95). No symptoms appeared at 15°C but 70% test plants showed symptoms when plants maintained at 15°C for 23 days were transferred in green house for 22 days (93). The heavy out break of the disease in Thailand was attributed to increase in acreage of mungbean and unusually dry conditions which allowed a population build up of the vector, *B. tabaci* (278).

4.1.3. MUNGBEAN YELLOW MOSAIC VIRUS (Philippines strain):

The yellow mosaic of mungbean has been reported from Philippines causing considerable losses (25, 26). It causes necrotic local lesions and systemic yellow mosaic in mungbean, necrotic local lesions on cotyledonary leaves of *Vigna unguiculata* and vein necrosis of *Phaseolus vulgaris*. The virus is isometric, measuring 25-30 nm in diameter. It is highlry stable *in vitro* being infectious even at 100°C after 10 minutes exposure and at 10^{-8} dilution. When stored in crude extract the virus lost its infectivity after 3 weeks at room temperature (25-32°C) but remained infectious for more than 3 months when kept at freezing temperature. It is a seed borne virus and readily sap transmitted but not apparently transmitted by aphids, leaf hoppers (*Nisia atrovenosa* and *Sogatella furcifera*) or by *Bemisia tabaci* (76). Plants exhibit bright yellow patches which enlarge and coalesce making the entire leaf chlorotic but leaving the main vein green (26). Sap inoculation causes local necrotic lesions while succeeding uninoculated leaves show systemic yellow mosaic. *V. unguiculata* is local lesion host (26).

4.1.4. BEAN YELLOW MOSAIC VIRUS :

The Bean Yellow Mosaic Virus is reported to infect mungbean in Indonesia (99) and blackgram in India (81). It induces the mosaic type disease in both the crops. When greengram plants are inoculated,it does not show any symptom on inoculated leaves but mosaic symptoms appear on newly comming un-inoculated leaves. The virus infects blackgram systemically. The infected plants show retarded growth, bear fruit much later with less yield as compared to healthy plants. The virus reduces the number and weight of mungbean seeds considerably (99).

Virus :

Bean Yellow Mosaic Virus (BYMV)

The virus obtained from mungbean has been studied in Indonesia (99). It consists of elongated flexuous particles 700-750 nm in length. The thermal inactivation point of the virus in crude juice of soybean leaves is 55°-60° C for 10 minutes and the dilution end point is 1: 10000 - 1:1000000. The longevity *in vitro* is 3-7 days at room temperature (24-30°C). Serologically the virus is found related with the Bean Yellow Mosaic Virus.

Transmission : The BYMV usually is not seed borne but it is found transmitted through mungbean seeds (Siwalik). The transmission per cent is very low (about 1%). The virus is transmitted mechanically as well as by *Aphis craccivora* and *Aphis glycines* from sovbean and mungbean to soybean and cowpea (99).

Host range : Thirty six plant species of 11 families were tested of which 21 plant species belonging to 6 families were susceptible (99). It indicates that the virus has very wide host range than MMV. *Chenopodium amaranticolor* is the local lesion host of the virus.

Host parasite interaction :

In blackgram variety Sindkheda 1-1, virus decreased the chlorophyll and starch. Sugars increased at time of symptom appearance but decreased later (81).

Control :

Use disease free seed. Resistant varieties not reported.

4.1.5. ALFALFA MOSAIC VIRUS (AMV):

Mungbean was found infected by Alfalfa virus 2 as early as 1934 (177). In Iran, mungbean plants infected with Alfalfa Mosaic Virus were found scattered in experimental crop of mungbean (107). Yellow mosaic of the foliage was the predominant symptom of Alfalfa Mosaic Virus on mungbean. The virus is readily transmissible by sap to a very wide host range which includes over 305 plant species belonging to 47 dicotyledonous families (173). The virus is transmitted by more than 10 species of aphids in non-persistant manner. Lucern is the main reservoir of the virus which should be avoided adjacent to mungbean crop.

4.1.6. HORSEGRAM YELLOW MOSAIC VIRUS (HYMV) :

The Horse Gram Yellow Mosaic Virus has been reported to infect several species of *Phaseolus* in India including *Phaseolus aureus* and *P. mungo* (136). The virus has been detected in naturally infected plants of greengram showing yellow mosaic symptoms. The virus is not transmitted mechanically (134).

4.2. VIRUSES CAUSING MOSAIC/MOTTLING SYMPTOMS :

Several viruses have been reported to cause mosaic and mottling symptoms in greengram and blackgram but only few are described in detail. The information on many of them is very inadequate. The symptoms caused by these viruses may differ from each other to variable extent but most of them produce systemic mosaic/mottling symptoms predominantly. The detailed study on mosaic causing viruses of blackgram was initiated mainly after 1960, though, before it diseases have been occurring on these crops (18, 69, 177).

4.2.1. BEAN COMMON MOSAIC VIRUS (BCMV) :

The bean mosaic virus was reported to infect artificially the mungbean (177) and urdbean (158) crops quite early. The virus was transmitted to both the crops form *P. vulgaris* (19). Shahare and Raychaudhury (208) were the first to observe the disease on urdbean under natural conditions in India. They described the disease in detail under the name mosaic of urdbean and identified the cause to be a strain of Common Bean Mosaic Virus. The disease was later observed wide spread in western Uttar Pradesh, India and was renamed as 'mosaic mottle' on the basis of its characteristic symptoms (161). The cause of the disease was confirmed as Bean Common Mosaic Virus (259, 260). The disease occurs in Iran (106) and Indonesia (99) where it is described under the name Mungbean Mosaic Virus (MMV). The causal virus infects mungbean as well as urdbean and occurs in several South-east Asian countries including Philippines and Thailand (24,284). The disease is reported to occur in USSR also (279).

Nature and extent of damage :

The disease is reported to cause 69% yield losses in blackgram plants showing symptoms at first trifoliate leaf stage (161, 264). The viability of pollens and seed, colour of seed, their germinability and number of seeds per pod are not affected adversely (4). The reduction in yield is contributed mainly by reduced number of pods/ plant and seed weight (4, 161). The inflorescence is converted in to leafy structures causing complete losses in some plants which may range up to 30 per cent (4). The yield of mungbean plants, raised from infected seeds, may be reduced by 31-75 % depending on different varieties. The seeds from the infected plants are discoloured, deformed and smaller in size with erratic and reduced germination (106). In Indonesia, the number and weight of seed of mungbean variety Bhakti were found to reduce by 64-78% due to the infection by Mungbean Mosaic Virus (99). However, it was not clear that whether this reduction was due to the virus or the susceptibility of the variety to the virus.

Symptoms :

The disease is characterized by a mosaic pattern of broad patches of light and dark green areas, blistering of the leaf blade, reduction of plant growth and shedding of flowers (208).

Symptoms of the disease have been described in detail on urdbean (161) and they appear in the form of light green patches alternating with normal green areas on leaves (Fig. 4.2). The leaves unfolded after infection are deformed, reduced in size and with margins rolling upward. Later on these young leaves show puckering and blistering with normal green areas present on both sides of larger veins only. The leaves unfolded still later tend to be assymmetrical,

Fig.4.2.
Mosaic mottle of green gram.

wrinkled and twisted. The leaves become rough and brittle with the age and plant remains stunted with excessive branching. The inflorescence, partly or wholly, is changed in to leafy structures depending on whether the infection is late or early. In severe cases, complete yield losses may be observed. The symptoms vary widely according to the variety of host, climatic conditions, time of infection and several other factors. The diagnostic characteristic symptoms of the disease are conspicuous reduction in size followed by typical mosaic mottle of trifoliates, partial or complete modification of the inflorescence in to vegetative structures and stunting and bushy appearance of the infected plant in the field (261, 262). Late infected plants may produce pods which are less in number, curved, misshaped, poorly filled or empty and are not produced at the same time. The symptoms on plants produced from the infected seeds are similar to those produced by sap transmission in the field except that the primary leaves show slight chlorosis, mottling and distortion along with downward cupping of leaves under seed transmission (259).

Symptoms on mungbean are produced on primary leaves within 3-5 days of inoculation. Subsequently, the disease becomes systemic producing, mosaic, deformation, puckering, rolling and blistering on foliage and stunting of plants (106). The symptoms of BCMV on mungbean are very much similar to those produced by Cucumber Mosaic Virus and may be confused with each other (106). However, CMV is not seed borne.

Virus :

Mungbean strain of Bean Common Mosaic Virus (M-BCMV).
Synonym : Mung Mosaic Virus (MMV)

The Bean Common Mosaic Virus causes the mosaic disease of blackgram (208) as well as greengram (106). However, the descriptions of BCMV isolates infecting these crops indicate considerable variability.

Morphology : In India, the urdbean isolate of BCMV was shown to pass through Chamberland candle L1 (208). The virus is flexuous and filamentous with a particle size of 700-750 nm in length (24, 99, 106, 286).

Thermal inactivation point : The reports regarding the TIP of the BCMV of mung/urdbean differ considerably from different places. The TIP in crude juice with an exposure of 10 minutes was reported 58°C (55-60°C) for urdbean isolates in India (208, 243) while it was 60-70°C for isolate studied in Philippines (24). The isolate of the virus from mungbean exhibited the TIP 55-60°C in Indonesia (99) and 60-65°C in Iran (106).

Dilution end point : The DEP of the virus by most of the workers has been reported to be 10^{-3} to 10^{-4} except Shahare and Raychaudhury (208) who reported it 1 : 2000 for blackgram isolate in India.

Longevity : The longevity depends on the temperature at which the virus is stored. In case of blackgram isolate it is reported 28h at room temperature (208) and 36-48 h at 24°C (243). The mungbean isolates have been reported to survive *in vitro* for 3-7 days in Indonesia (99) and 14-17 days (at 20°C) in Iran (106).

Serology : Serologically the blackgram isolate was found related to

Bean Common Mosaic Virus in India (243). The mungbean isolate of Thailand reacted strongly with the antiserum of BCMV isolate from bean in Japan (286). In Iran, the mungbean isolate was found serologically related with the Dutch isolate of BCMV (106). In Haryana, India, a virus causing mosaic of mungbean and urdbean and designated as MUMV-1 was reported serologically related to BCMV (241).

Transmission : It is transmitted mechanically, through seed as well as vectors.

(i) **Mechanical** : The blackgram as well as greengram isolates are transmitted mechanically by juice inoculation. The initial symptoms develop within 8-10 days after inoculation (208). The inoculum diluted to 1:100 results in quicker infection as compared to 1:1 extract (161). The sap transmission of urdbean isolate is reported 66.1 and 28% in urdbean and mungbean respectively (264).

(ii) **Seed transmission** : The virus is seedtransmitted in urdbean to an extent of 3-12% (161, 208). The presence of the virus has been detected in seed coat, cotyledons and primary axis of urdbean (5, 6). The extent of seed transmission depends on the age of the plant when it is infected. The seeds from plants showing symptoms 5-30 days after sowing, exhibited 17.3% transmission while it was only 5.1% in seeds collected from plants which were infected 40-50 days after sowing. The virus is seed borne in case of mungbean also to an extent of 8-32% (99, 106) in different varieties. Agrawal *et al.* found 2-14% seed transmission in different lines of mungbean (3). Plants infected after flowering do not give infected seeds (173). The small seeds show more transmission than shrivelled and bold seeds. The virus is not eliminated from the seed by dry heat or hot water treatment (7).

(iii) **Vector transmission** : In blackgram, the virus is transmitted by *Aphis craccivora* and *Aphis gossypii* to an extent of 75 and 69% respectively (243). It is transmitted by trasonemid mite, *Polyphagotarsonemus latus* (Banks) also which is the first record of tranmission of a plant virus by this mite (29). In Philippines, *Rhopalosiphum maidis* has also been reported to transmit BCMV in blackgram (24). In mungbean the M-BCMV is transmitted by *A. craccivora, Acyrthosiphon pisum* and *Acyrthosiphon sesbanae* in non persistant manner (106). The virus is acquired within 2 minutes or less feeding probs and the transmission is faster in Bountiful beans as compared to mungbean.

In Indonesia, the Mungbean Mosaic Virus (MMV) was transmitted in cowpea by *Aphis craccivora* and *Aphis glycines* to an extent of 70 and 83% respectively when 1 h pre-acquisition fasting, 15 minutes acquisition feeding and 10 minutes inoculation feeding periods were allowed (99).

Host range : Out of several plant species tested, *Phaseolus vulgaris, P. aconitifolius, P. aureus* and *P. mungo* were found susceptible to urdbean isolate of BCMV in India (161, 208, 243). Cultivars of *P. vulgaris* differed in showing symptoms. Some exhibited systemic mottling while others showed ring spot or necrotic local lesions. *Cajanus cajan, Glycine max, Pisum sativum, Arachis hypogaea, Lathyrus odoratus, Vicia faba, Vigna sinensis, Dolichos biflorus* and several others were not infected. On the other hand the Philippine isolate of BCMV from urdbean infected *V. mungo, V. radiata, V. sinensis, Chenopodium amaranticolor* and *C. quinosa* (24). In Iran, the M-BCMV from mungbean infected mainly legumes except *C. amaranticolor* and *Nicotiana clevelandii*. It produced systemic symptoms in *P. vulgaris, P. aureus, Lens culinaris,* fenugreek and *Vicia narbonensis* (106). In Indonesia, 37 plant species were tested of which *C. amaranticolor, P. vulgaris, V. faba, Vigna sinensis* and *Sesamum indicum* exhibited local lesions (99). The last three hosts, later, dveloped the systemic mosaic symptoms also. Inoculation of *Phaseolus angularis, P. radiatus,* and *Vigna sesquipedalis* exhibited no symptoms on inoculated leaves but systemic mosaic symptoms developed on newly developing leaves. No symptoms developed on *Beta vulgaris, Gomphrena globosa, Glycine max, Lathyrus odoratus P. vulgaris* and *Pisum sativum* but the virus could be dtected in leaves by back inoculation tests.

Host parasite interaction :

The palisade cells of infected urdbean leaves are found to be arranged loosely and irregularly. The spongy cells are less numerous each containing fewer chloroplasts as compared to those in healthy leaves (208).

Infection by BCMV interferes in the host metabolism affecting the amino acid contents of mungbean leaves (64). It decreases the activity of nitrite reductase in the leaves considerably (63). This decrease is reported more in plants grown from seeds without *Rhizobium* inoculation as compared to those from inoculated seeds (63). The infection also affects nodulation adversely but increases the dry matter contents (86, 247). It increases total amonical nitrogen, carbohydrate fraction, organic carbon, C : N ratio, leghaemoglobin and the bacterial population in nodules (65, 247). It also decreases the nitrogenase activity in the root nodules (65).

The activity of BCMV in mungbean is increased when seed is treated with *Rhizobium* or the nitrogen is supplied to plants. It increases the severity of symptoms but they apear earlier in plants from uninoculated seeds as compared to those inoculated with the *Rhizobium* (253).

Disease cycle and epidemiology :

The incidence of mosaic of mungbean and urdbean is observed more in crops grown during the rainy season as compared to that of spring season crop in Haryana, India (288). In the field, the first occurrence of the disease is seed borne. Further spread of the disease is through insects and is influenced by several factors. The symptom expression and the incubation period of BCMV in urdbean is affected by the stage of the plants, inoculum potential, salt concentration and pH of the buffer used for extraction of the infective sap and the temperature during and after the inoculation of plants. The optimum conditions favouring the fullest expression of symptoms and the shorter incubation period are: inoculation of the test plant at third trifoliate leaf stage with standard inoculum of 10⁻ dilution extracted in potassium phosphate buffer of 7.6 pH and 0.2 molarity (261). Placing the inoculated plants at 22-26°C results in early appearance of severe symptoms.

Host resistance :

Nene (161) reported only one urdbean variety, T 27, showing mild disease reaction as compared to other varieties. The urd bean entries A 6, PLU 584, T 27 and 1-4 - 2-2 have been reported to be tolerant (264). Out of 14 mungbean varieties tested, only T 2 and Hy 45 were found free from the disease (161). In Iran, 111 accessions of mungbean from 15 countries were tested of which only 2 were found free (Hy 45, accession 10855) from mungbean mosaic (106). In India, Singh, *et al.* (264) tested several mungbean varieties and reported L 293 and L 55-1 to be resistant. No correlation was observed between the field susceptibility and response of varieties to sap inoculation. The entries were scored as per scale given below:

Grade	Reaction	Symptoms
0	Nil	Symptoms not produced.
1	Light	Faint mosaic mottling with no signs of puckering and blistering of lamina in 50 % or less plants.
2	Mild	Above mentioned symptoms in over 50 % plants.
3	Moderate	Conspicuous mosaic mottling with slight puckering and blistering of lamina in 50 % or more plants.
4	Less severe	Above mentioned symptoms in over 50 % plants.
5	Severe	Mosaic mottling, puckering, blistering and distortion of leaf lamina.

Kaiser *et al.* (107) rated the mungbean entries against Mung bean Mosaic Virus as per 1 to 10 point scale in which 1 = no disease and 10 = all the plants severely infected.

Control :

Seed transmission of the disease is common and the important source of initial crop infection. Hence, clean and healthy seed procured from the healthy crop should be used. The possible time of sowing may be adjusted to coincide with the low aphid population. Control of insect vectors by using proper insecticide at proper time will help in reducing the spread of the disease in the field. Care should be taken to avoid contact of healthy and diseased plants during the cultural operations.

.2.2. MOSAIC DISEASE OF MUNGBEAN :

A mosaic disease of mungbean was reported from Tamil Nadu, India in 1966 the cause of which was tentatively identified as the leguminous strain of Tobacco Mosaic Virus (187).

Symptoms :

In the field, the plants remain stunted with reduced size of internodes and smaller leaves with mosaic mottling of light green to pale yellow patches interspersed with dark green areas. On inoculation, faint vein clearing of young leaves followed by small chlorotic spots is observed. The spots later coalesce to form light green areas alternating with dark green patches. In severe cases, the leaves are reduced in size, internodes shortened and plants remain stunted.

Virus :

The virus was tentatively identified as a legume strain of Tobacco Mosaic (TMV) with the thermal inactivation point more than 90°C when exposed for 10 minutes; dilution end point not detected up to a dilution of 1 : 1000000 and the longevity *in vitro* for more than 4 days at 5°C (187). The virus was transmitted by sap but not by species of aphids tested. Out of 22 plant species, *Vicia faba, Phaseolus vulgaris, Dolichos biflorus, Chenopodium murale, Nicotiana companiculata, Petunia hybrida, Nicotiana rustica* and *Chenopodium amaranticolor* were susceptible to it while *Vigna sinensis, Cajanus cajan, Cicer arietinum Pisum sativum* and *Crotalaria juncea* were not infected.

Host parasite interaction :

Infection of mungbean by a strain of TMV is reported to increase the ribosome content with which the presumptive virus messenger RNA was associated (130). In the mosaic affected plants of greengram, the starch contents are not affected but the sugars are reduced greatly. It does not affect the phosphorylase and amylase activity in mungbean plants (295). Washings of mosaic infected leaves of greengram contain more amino acids and reducing sugar as compared to un-infected leaves (171). It influences the quantity as well as quality of phyllosphere micro-flora.

4.2.3. DOLICHOS ENATION MOSAIC VIRUS (DEMV) :

Greengram and blackgram are infected by the Dolichos Enation Mosaic Virus which is a leguminous strain of Tobacco Mosaic Virus (TMV). It was reported from Tamil Nadu reducing the yield i blackgram and greengram to an extent of 64.3 and 79.7 % respectively (189.190).

Symptoms :

The inoculation with DEMV induces elongation of the main axis as well as branches of the inflorescence. The glands at the base of the flowers enlarge and their number is increased. No flower is formed on these gland points but few are formed on the tips of branches of inflorescence (191). Further commencement as well as attainment of 50 % flowering is delayed by 4-10 days. Flowers do not drop and the pollen sterility does not occur. Besides elongation of inflorescence, proliferation in the numbers and size of nectaries takes place. The yield is drastically reduced due to reduction in flowers, pods/plant and seeds/pod.

Virus :

Dolichos Enation Mosaic Virus is a leguminous strain of Tobacco Mosaic Virus (TMV). It is transmitted by sap but no virus has been detected in seeds. It is transmitted by the Slao spider mite (*Tetranycnus ludeni*) easily to several plant species including mungbean and urdbean (182). Cluster bean is the local lesion host of DEMV. Though, the virus is not seed transmitted, the seeds from the virosed plants are small and deformed with reduced germination (182). The virus has been detected in the nodules of infected plant with slight increase in the nitrogen contents (190).

Control :

No specific control measures are recommended but it is reported that the disease syndrome is more marked in winter as compared to summer crop. The blackgram variety Mash 69 and the greengram Pusa Baisakhi are tolerant (190).

4.2.4. CUCUMBER MOSAIC VIRUS (CMV) :

A strain of Cucumber Mosaic Virus causing mosaic disease of brinjal was reported to infect greengram and blackgram producing local lesion symptoms in India (207). The natural occurrence of this virus on greengram was first reported from Iran (107). The virus has been reported to infect mungbean in Columbia (178), Indonesia (99),

Japan and Taiwan (98). It affects the plant growth adversely reducing height, branch length and the dry weight to a greater extent (180). The yield of susceptible mungbean varieties may reduce more than 50 % due to the infection by the virus (266).

Symptoms :

The brinjal isolate of the virus produces local lesions in greengram as well as blackgram (207). In Iran, CMV is reported to cause stunting of mungbean plants with deformation and curling of leaves (107). These symptoms resemble to those caused by the Mungbean Mosaic Virus which is a strain of Bean Common Mosaic Virus. On the susceptible mungbean variety, the virus produces mosaic, veinal chlorosis or necrosis, leaf distortion, stunting, flower abortion and pod distortion (181). The symptoms produced by the Cucumber Mosaic Virus on mungbean and other legumes are very much variable (173).

Virus :

Cucumber Mosaic Virus mungbean strain (CMV-M)

The virus CMV, isolated from mungbean has been studied in detail in regard to its morphology, properties, transmission and host range characters (181). It has isometric particles of 30 nm size with TIP 55-60°C, Dilution end point 1:500 to 1:1000 and longevity at 22-24°C for 48 h. The CMV isolated from mungbean varieties received from Taiwan (98) is reported to have isometric particles with 28 nm size showing positive serological reactions with the legume strain of Cucumber Mosaic Virus (CMV-L). Based on the studies made on its properties, host range, symptoms etc. this isolate has been identified as mungbean strain of Cucumber Mosaic Virus (CMV-M).

Transmission : The virus strain studied by Purivirojkul *et al.* (181) is sap transmitted and is seed borne to an extent of 0.61 per cent. It is transmitted by *Aphis craccivora* in a non persistant manner up to an extent of 40 % if 8 aphids per plant are used. Probably the virus is limited to the stylet only. The isolate studied by Iwaki (98) is also seed transmitted to an extent of 1.1 % and that it is transmitted in susceptible mungbean varieties by *Myzus persicae* to an extent of 50 per cent.

Host range : The virus is poorly transported in cucumber, tomato, and tobacco but is able to infect 23 host plants belonging to families Amaranthaceae, Chenopodiaceae, Cucurbitaceae, Solanaceae and Leguminosae (181). The isolate studied by Iwaki (98) produces various types of symptoms on different host plants and resembles with that described by Poehlman (178) who designated the virus as CMV-M.

The infection of the virus results in the quantitative changes in the peroxidase activity in mungbean plants which is different from that induced by the senescence (170).

Host resistance :

Sources of resistance against the CMV have been identified in mungbean. The entry PI 271401 is highly resistant (181). Many of the cultivars of mungbean are not invaded by the virus systemically which results in no seed transmission of the virus and thus causes negligible damage (98). The resistance in PI 271401 is conditioned by a single dominant gene designated as Cmm. Its locus is not linked with colour of the seed, texture, layer or colour of the seed coat (266).

Control :

Method of detection of CMV-M in the seed has been described (176) which may help in selecting disease free seed for use in sowing. It will help in minimizing the disease.

The control of vector by using systemic insecticide may protect the crop from the virus slightly (180).

4.2.5. GREENGRAM MOSAIC VIRUS :

A mosaic disease of greengram was observed occurring in Tamil Nadu, India which was found different from those reported earlier on this crop. The disease was characterized by the mosaic mottling symptoms on the leaves. The virus was studied in detail and designated as the Greengram Mosaic Virus (185).

Symptoms :

No symptoms are observed on inoculated cotyledonary leaves but typical mosaic mottling is seen on the subsequent leaves. Symptoms produced by all the 4 isolates of the virus (GGM 1 to 4) differ from each other in the field.

GGM 1 : Vein banding and mosaic mottling with dark green areas of raised blisters and slightly stunted plants. On inoculation, chlorotic ring spots followed by mosaic mottling with dark green areas interspersed with light green patches and reduced leaf size are observed.

GGM 2 : In field, severe mosaic mottling with dark green areas and yellow spots on lamina are observed. On inoculation, vein clearing followed by mosaic mottling is developed.

GGM 3 : In field, leaves show elongation with severe puckering, crinkling and clustering appearance. On inoculation, mild mosaic develops on young leaves.

GGM 4 : In the field, vein clearing followed by mosaic mottling with dark green raised blisters are seen. On inoculation, it produces mosaic mottling of dark green and light green patches.

Virus :

Greengram Mosaic Virus (GGM 1, GGM 2, GGM 3, GGM 4)

The GGM has been reported to differ from the TMV causing mosaic disease of greengram (184). The 4 isolates of the virus differed slightly from each other in their properties. The GGM 1 to 3 were not inactivated even at 90°C when exposed for 10 minutes but isolate GGM 4 was inactivated at 85°C. The Dilution end point was 1 : 100000 to 1 : 1000000 in case of GGM 2 to 4 but it was ! : 10000 to 1 : 100000 for GGM 1. When stored in sap at room temperature (28-30°C), the GGM 2 to 4 lost their infectivity after 96 h while GGM 1 remained infective for 120 h. All the isolates were sap transmissible but not by seed. The GGM 1, 2, 3 were transmitted by *Aphis gossypii* and *Aphis craccivora* but GGM 4 was transmitted by *Myzus persicae* only. Serologically all the isolates were related with each other but not with TMV, Potato virus X, cowpea mosaic and soybean mosaic.

Host range : Out of 48 plant species tested, only 8 were infected. They were *Cyamopsis tetragonoloba, Medicago sativa, Phaseolus vulgaris, Pisum sativum, Physalis floridana* and *Solanum nigrum* with *Phaseolus aureus* and *P. mungo*. The isolates GGM 2 and 3 reacted identically on these plants while GGM 1 and 4 showed different reaction. *P. aureus* was infected by all the 4 isolates but *P. mungo* was infected by GGM 2, 3, 4 only. On *P. vulgaris* and *C. tetragonoloba* only local lesions were produced. The isolates GGM 2 and GGM3 were similar in regard of their physical properties host range etc.

Control :

No work has been done but control of aphid vector may help in reducing the disease severity.

4.2.6. GREEN MOSAIC VIRUS OF MUNGBEAN :

This virus has been reported from Philippines (25). It does not induce necrotic local lesions in mungbean and *Vigna unguiculata* but produces numerous, small, faint, chlorotic local lesions on *Chenopodium amaranticolor*. In mungbean, inoculated cotyledonary leaves usually exhibit vein necrosis. The uninoculated first trifoliate leaves show vein yellowing with several tiny yellow dots followed eventually by systemic green mosaic. The virus particles are rod shaped, 333 nm in size. It is transmitted by seed, sap and aphids and is inactivated at 90-100°C. The DEP is between 1 : 10^4 - 10^5.

4.2.7. MUNGBEAN SEVERE MOSAIC VIRUS (MBSMV) :

Several viruses have been reported to produce mosaic symptoms in mungbean and the Mungbean Severe Mosaic Virus is one of them causing losses in India (250). The infection of mungbean by the virus alters the metabolism of the host plant ultimately reducing the yield. It reduces the chlorophyll contents and rates of net and gross production but the respiratory losses are increased in the infected leaves of greengram (251). The sugars and the starch contents decrease in the infected leaves, stems and roots. In diseased fruits, the total sugars, reducing sugars, non reducing sugars, total nitrogen, protein, nitrate nitrogen, total free amino acids and the inorganic phosphorus are detected higher as compared to their healthy counterparts. However, starch, ammonical nitrogen organic phosphorus and the total P show the reverse trend (252). The activities of the peroxidase and catalase have been reported to be higher and lower respectively in the infected mungbean leaves as compared to healthy leaves. These activities increase up to the 80th day after inoculation and then decline. The severity of symptoms also increasesup to 80th day and then become moderate (250). This virus differs from other viruses infecting mungbean and hence named as Mungbean Severe Mosaic Virus (MBSMV).

The mungbean mosaic has been reported to be reduced by controlling the aphids through spraying the insecticide in Philippines (67).

4.2.8. URDBEAN MOSAIC VIRUS :

This disease was observed on urdbean during 1970-72 near Gorakhpur, Uttar Pradesh, India (249). The disease is not wide spread but it is able to attack urdbean as well as mungbean crops.

Nature and extent of damage :

The disease cuases serious reduction in the yield of urdbean. The reducton depends on the stage at which the plants are infected. It is reported to an extent of 85, 62 and 43% in cases of early, mid and late inoculations respectively. The reduction in yield is contributed by reduction in number of pods/plants, seeds/pod and the seed weight (223).

Symptoms :

Leaves of the affected plant show mottling, vein banding, malformation and reduction in size. Plants grow poorly and bear only few pods. Inoculation of urdbean by the virus results in vein clearing followedby appearance of interveinal light green areas. In advanced stages, leaflets become chlorotic with prominent vein banding along the main vein. Plants remain stunted with smaller leaves and pods. Only few pods are produced which often show light green to whitish green spots on their surface. On mungbean, similar symptoms are produced but at advanced stages mosaic symptoms and leaf distortion are seen. The growth of such plants is greatly reduced (249).

Virus :

Urdbean Mosaic Virus (Singh and Singh 1978)

The physical properties of the virus are some what similar to cowpea strain of Southern Bean Mosaic Virus (SBMV-C) but the cross protection tests suggest that it is quite distinct from SBMV (249). It does not resemble the Cowpea Mosaic Virus, urid mosaic virus (BCMV), Blackgram virus and SBMV infecting urdbean.

Thermal inactivation point : The TIP of the virus is between 95-100°C when exposed for 10 minutes.

Dilution end point : The Virus looses its infectivity at 10^{-6} to 10^{-7} dilution.

Longevity : The longevity *in vitro* when stored at 5-7°C is 35-38 days. The virus remains active in leaves of urdbean desiccated over calcium chloride for more than 950 days at room temperature (12-40°C). *In vitro* tests the multiplication and infectivity of the virus is reduced by milk of goat, cow and buffalo to an extent of 90, 50 and 70% respectively (225).

Transmission : The virus is easily transmitted mechanically by sap and through seed (249). *Aphis craccivora, A. gossypii, A. nerii, Myzus persicae* and *Rhopalosiphum maidis* are unable to transmit the virus.

Host parasite interaction :

Infection by urdbean mosaic virus influences the metabolism of the host greatly affecting carbohydrate and phosphorus contents (246). It decreases chlorophyll contents and catalase activity but increases peroxidase activity in the leaves (245). The infection also increases the nitrogenous fractions except amonical nitrogen (total nitrate and nitrite nitrogen, total free amino acids and protein) of the fruit parts of urdbean (223). Carbohydrate fractions are increased but phosphorus fractions are decreased.

The infection causes reduction in the activity of catalase and nitrogenase but increases the peroxidase and nitrate reductase activity in urdbean nodules (224). It increases total nitrogen, nitrite and nitrate nitrogen, protein and total free amino acids but decreases the amonical nitrogen in nodules of blackgram (227). Infection also reduces the leghaemoglobin content of blackgram nodules (26).

Disease cycle and epidemiology :

The first infection in the field is probably seed borne and the further spread seems to be due to the contact. The virus produces mild symptoms in winter when the temperature ranges 12-15°C as compared to that in summer (22.5-40°C).

Host resistance :

The cultivars vary in their susceptibility (245). All the varieties of mungbean including T 44, T 51, Hyb 45, Jalgaon 74, B 1, Khargon and Krishna and urdbean T 9, T 27, T 65. L 35-5, Mash 1, Mash 48, No 55, L 41-13, Co 1, Khargon and Krishna were infected but mungbean T 44, T 51, Khargon and urdbean T 9, T 27, T 65 and mash 1 were most susceptible (249).

Control :

The seed should be procured from the disease free crop. The disease is inhibited by spraying plants with milk of goat just before or after the inoculation (225). Foliar spray of the extract of *Mirabilis jalapa* leaves causes marked suppression of symptoms improving growth and flowering and reduces the virus multiplication rate in urdbean plant (291).

4.2.9. SOUTHERN BEAN MOSAIC VIRUS :

The Southern Bean Mosaic Virus (SBMV) has been reported to cause mosaic and mottle disease in blackgram (248) and greengram (173) in India. It is reported occurring on mungbean in Indonesia also (9). Singh and Singh (248) isolated three strains of the virus from urdbean in India which had nearly the same host range. The hosts were restricted to Leguminosae only. The virus particles are isometric with 28 nm size. The virus shows the dilution end point 10^{-3} to 10^{-4}, the thermal inactivation point 70-80°C and longevity *in vitro* at 5-7°C, 65 to 72 days. It remains infective for 426 to 634 days at 12-40°C in leaves stored over $CaCl_2$ in desiccator. The virus is readily transmitted by sap but not by seed or aphids (248). The infection by SBMV increases total N, protein N, ash value in leaf and leaf protein concentrate, fibre content, lipid and phosphorus but decreases total soluble sugars, starch and calorific value. Free amino acids are increased but enzymetic hydrolysis decreases in infected plant (222).

4.2.10. MOSAIC DISEASE OF BLACKGRAM (Srivastava *et al.* 1969):

Some of the blackgram plants were observed near Lucknow, India which were affected by some virus (271). The main symptoms observed are mosaic mottling associated with reduction in size and narrowing of leaves. The diseased plants are stunted and the yield is greatly reduced. It produces mild mosaic symptoms on greengram on inoculation.

Virus :

The virus is transmitted mechanically but *Myzus persicae* from tobacco and *Aphis gossypii* from cowpea do not transmit the virus. It has the dilution end point 1×10^{-5}, thermal inactivation point, above 90°C when exposed for 10 minutes and longevity more than 8 days at 30-33°C. The virus differs from the BCMV reported by Shahare and Raychaudhury (208) in its physical properties and host range and also from other sap transmissible viruses reported on blackgram so far.

The virus is transmitted to *Cyamopsis tetragonoloba, Dolichos biflorus, Phaseolus mungo* and *P. aureus.* In blackgram, uninoculated leaves develop severe mosaic while the young developing leaves remain smaller and narrow and less pods develop on affected plant. On mungbean, mild mosaic symptoms develop on young leaves after 7-8 days of inoculation while necrotic local lesions are formed in case of *C. tetragonoloba* and *D. biflorus*.

Control :

No control measures are recommended but blackgram varieties T 65 and T 67 are resistant.

4.2.11. MOSAIC DISEASE OF BLACKGRAM (Ramakrishnan *et al.* 1973) :

Five virus isolates infecting blackgram were collected and studied in detail by Ramakrishnan *et al.* (186).

Symptoms :

The symptoms on blackgram produced by all the isolates differ from each other as follows :

BG 1 : Mosaic mottling of dark green areas interspersed with light green patches. The symptoms later mask.

BG 2 : Vein clearing of young leaves after 12-15 days of inoculation. Later, they show crinkling and puckering. On some young leaves, dark green raised blisters develop. Plants show stunting with proliferation of axillary buds.

BG 4 : Vein clearing and deformation of leaves and mosaic mottling with blistering.

BG 5 : Light green spots giving mottling appearance with elongation and curling of leaves.

B 7 : Vein clearing of young leaves after 10-15 days of inoculation. Later, systemic mosaic mottling, reduction in leaf size; crinkling and deformation of leaves with reduction in growth of severely infected plants.

Virus :

Properties of the 5 isolates are complied in the Table 4.2.

Table 4.2

Properties of 5 isolates of the blackgram mosaic virus.

Isolate	Dilutien end point	Thermal inact-tivation point (°C)	Longevity *in vitro* (h) at 28-30°C
BG 1	1: 10000-1:100000	55-60	72
BG 2	1: 1000 -1: 10000	50-55	32
BG 4	1: 1000 -1: 10000	50-55	80
BG 5	1: 100000 -1:1000000	80-85	72
BG 7	1: 100000-1:1000000	85-90	96

Host range : Isolates BG 2 and BG 7 are similar in host range as they infected *Solanum nigrum* in addition of *P. mungo*. Isolate BG 1 infected *P. mungo* only while BG 5 could infect *Phaseolus aureus* in addition to *P. mungo*. Isolate BG 4 was able to infect *P. mungo, P. aureus, Glycine max, Vigna sinensis, Gomphrena globosa,* and *Nicandra physaloides.*

Transmission : All the isolates were transmitted mechanically but none by seed. Amongst the aphids, *Aphis gossypii* and *A. nerii* were unable to transmit these isolates but they were transmitted by *Myzus persicae* and BG 4 by *Aphis craccivora*.BG 2 and BG 7 were not transmitted by aphids tested. All the five isolates differ from each other in one or more of their characters. They also differ from the blackgram mosaic causing viruses reported earlier (208, 271).

Effect of blackgram mosaic on host :

Black gram mosaic reduces the height of the plant which is correlated positively with loss in grain weight. The weight and length of roots are not affected much but the number and weight of nodules are considerably reduced. There was a notable reduction in pod number and grain weight, A regression equation to predict yield losses, based on growth retardation, has been presented (151).

Control :

No control methods have been worked out so far.

4.2.12. BLACKGRAM MOTTLE VIRUS :

Blackgram plants infected with the blackgram mottle virus were observed near Delhi, India (176). The virus occurs on mungbean in Indonesia (201) and Thailand (94) under natural conditions.

Symptoms :

The symptoms of the disease have been studied in detail of which predominant is mottling in blackgram foliage (176). On other hosts the symptoms have been described by Scott and Phatak (205). It produces chlorotic lesions on primary leaves of mungbean which later become necrotic with mottling on secondary leaves. In Pinto bean, inoculated primary leaves show necrotic local lesions. In clusterbean, necrotic lesions are formed on the inoculated leaves without any systemic infection.

Virus :

Blackgram Mottle Virus (BlMV)

This virus has been studied in India, Thailand and Indonesia and all the isolates are more or less identical in regard of their properties (94, 176, 201, 205). The particles are isometric of 28 nm diameter with extinction coefficient 50, sedimentation coefficient 122 S, single electrophoratic component migrating towards the anode, virion density 1.364 g per cm^3 and 20 % nucleic acid. It has one nucleic acid component and one coat protein subunit with molecular weight of 1.4×10^6 and 38200 daltons respectively. The virus has TIP 90-92°C, DEP 5×10^{-4} to 1×10^{-5} and longevity *in vitro* for 40-45 days (176). The Thailand and Indonesian isolates are similar in this regard and have TIP 85-90°C for 10 minutes, DEP 10^{-9} to 10^{-10} and longevity *in vitro* 6-9 weeks at 20°C with UV light absorption spectrum typical of nucleoprotein components with a A 260/A280 values of 1.55 (94, 201). The Blackgram Mottle Virus has been found serologically related with the Broadbean Mottle Virus but differs from it and beetle transmitted viruses of legumes in some other properties (205).

Transmission : The virus is transmitted mechanically and by seed but not by aphids (176). It is transmitted by bean beetle (*Epilachna varivestis*) first day after acquisition but not after that (205). The Thailand isolate is transmitted by the beetle *Monolepta signata* (94) and the Indonesian isolate by *Colposcelis signata* and *Balisepta fulvipes* (201).

Host range : The host range of the Blackgram Mottle Virus is limited to legumes particularly cowpea, soybean, broadbean, mungbean, urdbean, pinto bean, black valetine bean (205). In Thailand, the virus has been isolated from several plant species belonging to families of Aizoaceae, Amaranthaceae Chinopodiaceae, Leguminosae, Pedaliaceae and Solanaceae (94). The virus is detected in the host by ELISA test and mechanical inoculation tests with whole seeds and at low concentration from seeds without coats (201).

Control :

No specific control methods have been recommended.

4.2.13. COWPEA MOSAIC VIRUS :

The greengram and blackgram were found to be infected by cowpea mosaic virus in Trinidad (70). In India, the cowpea mosaic disease was found to be transmitted to several hosts including blackgram and greengram producing systemic mosaic symptoms (157). At research farm Kanpur, 22-30% incidence of cowpea mosaic on blackgram was observed in year 1977 (20).

Symptoms :

The virus, under artificial inoculation conditions, produces chlorotic local lesions followed by yellow vein clearing of young developing leaves which are later accompanied by severe green and yellow mottling and blistering (70). The disease on the blackgram variety T 27 causes severe mosaic symptoms consisting of broad patterns of light and dark green areas. In severe cases yellowing of leaves and stunting of plants is observed. The disease is much conspicuous on young leaves as scattered yellowish green areas. specks and spots forming mosaic mottle pattern. The midribs and some prominent veins of the older leaves offen become twisted with their margins slightly wavy. The diseased plants mature late and bear fewer flowers and pods. The pods are smaller in size and bear smaller seeds remaining immature frequently (20).

Virus :

Cowpea Mosaic Virus

The virus has thermal inactivation point between 90-95°C, dilution end point 1 : 1000 to 1 : 10000 and longevity *in vitro* at 28-35°C between 48-72 hours. It is transmitted mechanically with aid of the carborundum and vectors *A. craccivora* and *A. gossypii* (20). It is transmitted by seed, *Myzus persicae and Aphis evonymi* also (157). The host range of the urdbean isolate is *D. lablab, Phaseolus acutifolius, P. calcaratus, G. max* and *V. sinensis* (20).

4.2.14. COWPEA MILD MOTTLE VIRUS :

The virus was isolated from the naturally infected black gram plants showing mottling and smalling of leaves in India (286). It infects *Chenopodium amaranticolor, G. max, P. vulgaris, V. mungo* and *V. sinensis.* On *C. amaranticolor.* the symptoms are diagnostic local lesions but diffused chlorotic lesions are observed on systemically infected leaves. Other hosts develop milk mottle and stunting of plants. Particles are 650 nm size filaments which break easily. Serologically it is related with cowpea mild mottle virus.

4.2.15. BLACKEYE COWPEA MOSAIC VIRUS :

The blackgram seeds collected from the infected plants in Iran were sown in Geneva. The resultant plants exhibited diffused green mottling of leaves, foliar cupping along the main vein and stunting of plant. The cause was identified as Blackeye Cowpea Mosaic Virus (BlCMV)on the basis of electron microscopy, host range, serology and no infection of cowpea line TVU 2480. The virus was seed transmitted up to 48% and was recovered from the embryo but not testae of seeds. Storage of seed for 3 years did not eliminate the virus (179).

4.2.16. ARHAR MOSAIC VIRUS :

The greengram plants infected with Arhar Mosaic Virus had lower dry weight and high moisture contents. It reduced the protein contents in leaf, stem and seed (243). The reduction in growth, fresh weight , dry weight was observed in mungbean as well as urdbean (244). It also decreased number, weight and size of root nodules in mungbean but increased the same in urdbean. When compared with the healthy plant, the total nitrogen was lowered in mungbean but it increased in urdbean plants.

4.2.17. MUNG URID MOSAIC VIRUS (MUMV 1 and MUMV 2) :

Various types of mosaic symptoms were observed on greengram and blackgram in Haryana, India with higher incidence in Kharif season as compared to summer season (288). The two causal viruses were idetified as Mung Urid Mosaic Virus 1 and 2 (MUMV 1 and MUMV 2). Both the viruses were transmitted by sap and seed but differed from each other in their host range, host reaction, physical properties and the serological reactions (241). The MUMV 1 was more related to BCMV serologically. The disease was more severe in mungbean as compared to urdbean (239). Both the viruses reduced the dry weight and length of root and shoot, leaf area, pod setting as well as yield of both the crops (238). The reduction in all the components was more in case of MUMV 1 as compared to MUMV 2. The yield reduction was 34-45% due to MUMV 1 and 10-28 % due to MUMV 2. Both the viruses caused reduction in chlorophyll `a' and `b', carotenoid and protein contents of leaves (240). The RNA and DNA contents were not affected much by MUMV 2 but MUMV 1 reduced them considerably in both the crops (241).

4.3. LEAF CRINKLE :

The urdbean leaf crinkle disease was first noticed in the states of Uttar Pradesh and Delhi, India in 1966 (298) and later in the Tarai region of Uttar Pradesh (159). The disease was characterized by stunted plants and crinkling and rugosity of leaves. The disease was suspected to be of virus origin but efforts of its mechanical transmission were not successful in the beginning (298). A disease, similar to leaf crinkle was reported from Punjab, India under the name "Curly top" which was transmitted by grafting but not mechanically 58). Later, the disease was found transmitted through sap, aphids and leaf hoppers (45,112). The disease is reported to be caused by Leaf Crinkle Virus, ULCV, (114, 115). In India, it has been reported from states of Haryana (239), Himachal Pradesh (89), Madhya Pradesh (2) and Andhra Pradesh. The disease occurs in Indonesia also (99) and has potential to cause heavy losses to urdbean crop, though, mungbean is also attacked.

Nature and extent of damage :

Urdbean Leaf Crinkle Virus reduces the plant height, root length, nodulation, pods/plant, length of pod, seeds/pod and 100 seed weight to an extent of 19.6, 5.3, 33.7, 84.0, 83.4, 27.0 and 1.6% respectively (60) but increases the dry weight of root and shoot and area of leaf (238). It results in serious yield losses which may range from 2.12 to 100 per cent depending on the time or the stage of the plant at which infection takes place (103, 118, 164). In urdbean, the yield losses due to ULCV are maximum when plants are infected 10 days after sowing and negligible at 60 days after sowing

(30). The losses in yield are contributed mainly by reduction in pods/plant and seeds/pod (30, 103, 118, 199).

Besides the reduction in the yield, the seed quality from the diseased plant is also affected adversely. The seeds become shrivelled and light brown in colour but 100 seed weight remains unaffected due to the production of some over sized seeds in addition to shrivelled ones (30). Though, most of the workers have observed no effect on the germination of seeds collected from the diseased plants (13, 30, 211), reduction of 46.6 and 26.0 % has been observed in germination of seeds collected from plants inoculated at the age of 10 and 25 days respectively (154). This effect seems to be due to the improper development of seeds under the influence of the disease and not due to the direct effect of the virus.

Symptoms :

According to the initial reports (159, 298) the urdbean plants affected by the disease are stunted with crinkling and rugosity of leaves. Khatri *et al.* (112) characterized the disease by leaf crinkling reduction in leaf size, witches broom and sterility. The symptoms of the disease have been studied in detail on naturally infected as well as artificially inoculated plants (116, 118, 161) which are described below-

In the field, first symptoms appear 3-4 weeks after sowing when third trifoliate leaf increases in size and becomes light green in colour. Typical leaf crinkling develops with in one week after the appearance of first symptoms. In severe infections, the stipules at the base of affected trifoliate become thicker and broader. The stipules at the base of two side leaflets become thicker and concave with pointed ends while those at the base of the terminal leaflet are rudimentary or absent. The petiole of the terminal leaflet becomes so short that the basal portion of its lamina closely touches the surface of both the side leaflets. Flowering is delayed and the peduncle of the inflorescence produced from the axil of the affected trifoliate is malformed having large number of small flower buds which are proliferated and most of them do not produce typical yellow petals. The sepals of buds become thicker, greener and cover the half or all of the buds giving bushy appearance to inflorescence. Pod setting is greatly reduced particularly in case of early infections.

When primary leaves of 6 day old plants are inoculated, the mild and transitory symptoms start appearing on second trifoliate leaf within 10 days. The third trifoliate increases in size with light green colour followed by crinkling on the upper surface which is evident 13 days after inoculation. Subsequent leaves produced after it also develop similar symptoms. Enlargement of leaves followed by crinkling of laminae are the characteristic symptoms of the disease (Fig. 4.3). With the further growth of the plant, the crinkling and rugosity on older leaves goes on decreasing but on younger leaves it becomes more pronounced and distinct. The tips of affected leaflets curve downward after a month of appearance of initial symptoms.

Fig.4.3.
Leaf crinkle of black gram.

Virus :

Urdbean Leaf Crinkle Virus (ULCV) Kolte and Nene 1970.
Synonym : Urd curly top (Chohan and Kaliha 1967).

Williams *et al.* (298) suspected urd bean leaf crinkle disease to be of virus origin but they could not transmit it mechanically. Kolte and Nene (115, 116) transmitted the disease mechanically and named the cause as Urdbean Leaf Crinkle Virus(ULCV).The cause can not be a viroid because its infectivity is not reduced by the phenol treatment (40). Narayanaswamy and Jaganathan (152) reported Blackgram Leaf Crinkle Virus (BLCV) from Tamil Nadu which slightly differs from ULCV in some of the properties. However, it may be treated as the synonym to ULCV (229).

Morphology : The partially purified ULCV is milky white in colour. The particles are isometric with an average diameter of 320 A` (77, 215).

Thermal inactivation point (TIP) : The ULCV is inactivated when exposed at 60° C to 70° C for 10 minutes (117, 152, 161). According to Dubey *et al.* (77) the TIP of ULCV lies between 60° and 75°C while Gupta (89) mentioned it between 70 and 75°C. In fact, the TIP of ULCV is pH dependent and its infectivity looses faster at pH 7.8 as compared to 7.0 at 70°C (40).

Dilution end point : The dilution at which the ULCV is unable to infect the host plant has been reported to be 10^{-4} to 10^{-5} (117, 161), 10^{-3} to 10^{-4} (89) and 1:5000 (152).

Longevity : The ULCV, in sap stored at room temperature (30-33°C), looses its infectivity on 4th day while that stored in refrigerator at 5°C becomes inactive on 9th day (117, 161). The longevity at room temperature has been mentioned to be 48 hours (152), 96 hours (89) and even 5 days (77). The longevity is influenced by several factors when ULCV is stored at room temperature. It increases from 3 days to 5 days in presence of 5% sucrose and up to 7 days if 1% mercaptoethanol is also added (40).

pH and infectivity : The inoculum of ULCV extracted in 0.01 to 0.07 M phosphate buffer,produces only mild symptoms while that extracted in 0.1 to 0.2 M phosphate buffer develops severe symptoms (161). The inoculum in 0.5 M phosphate buffer is unable to produce any symptom of the disease. The infectivity of ULCV is highly pH dependent and is maximum at pH 7.8 (40) and 7.2 (152). Any increase or decrease in pH affects the infectivity of ULCV adversely(59).

Other chemicals and infectivity : Several chemicals have been reported to influence the infectivity of ULCV greatly (40). The infectivity is increased by 25 % by adding 1-5 % sucrose and further by 33 % if mercaptoethanol (1%) is also added to the inoculum. Magnesium sulphate and magnesium chloride reduce the infectivity by 66 % while calcium tribasic and dibasic phosphate and EDTA completely inhibit the infectivity. Amongst the reducing agents, the mercaptoethanol increases the infectivity but sodium sulphate, ascorbic acid and cysteine decreases it considerably. As regards the antiviral chemicals, actinomycin D and cycloheximide decrease the infectivity *in vitro* as well as when sprayed on to plant immediately after inoculation with ULCV. Tetracycline hydrochloride is ineffective in this respect. Phenol reduces the infectivity of ULCV down to 20% indicating that it is not viroid because phenol acts by removing coat protein of the virus particle. The ULCV is insensitive to ether which indicates that it does not contain lipid material.

Serology : The ULCV is active serologially and precipitates up to 1:8 dilution of antiserum (152). Serologically it is not related with Bromovirus (3 isolates of Broadbean Mottle Virus), Potyvirus (2 isolates of Southern Bean Common Mosaic Virus and 2 isolates of its cowpea strain), Comovirus (3 isolates of Bean Pod Mottle Virus) and Blackgram Mottle Virus (77). ULCV is not related serologically with some members of Comovirus, Tymovirus, Bromovirus and Potyvirus groups (27).

Host range : Out of 32 plant species tested at Pantnagar, only mungbean (*Vigna radiata*), urdbean (*Vigna mungo*), mothbean (*Phaseolus aconitifolius*) and cowpea (*Vigna sinensis*) were infected by ULCV (161). In Himachal Pradesh, tobacco and *Phaseolus vulgaris* were also found susceptible to ULCV (89). Similarly in Tamil Nadu, *Cyamopsis tetragonoloba, Arachis hypogaea* and *Cajanus cajan* were found the additional hosts of Blackgram Leaf Crinkle Virus. The symptoms developed were narrowing of leaves and crinkling in *C. tetragonoloba,* cupping of leaves and chlorosis in *A. hypogaea* and crinkling of leaf margins in *C. cajan* (152). None of these species were infected when tested at Pantnagar (117, 161). Out of 60 plant species belonging to 9 families, *Cucumis sativus* L. var. National pickling and *Laginaria cylindrica* L. var. local were the additional hosts of ULCV showing malformed leaves with upward curling of margins and stunted plants with reduced tendency of trailing (28). Kadian (105) reported 5 weed hosts infected by ULCV with systemic infection in *Convolvulus arvensis.*

Transmission : The virus ULCV is transmitted through grafting, sap, seed and insects :

(i) **Grafting** : The cause of curly top of urdbean, now considered as synonym to ULCV, was first reported graft transmissible in Punjab (58). Later ULCV was shown to be transmitted by grafting by Gupta (89).

(ii) **Sap transmission** : ULCV is transmitted to urdbean when primary leaves of 6 day old plants are inoculated by sap from diseased plant with the help of carborundum powder. Symptoms of the disease appear 11-19 days after inoculation. The serial sap transmission has also been shown (116, 164).

(iii) **Seed transmission** : The ULCV has been reported seed transmitted to an extent of 18.4 per cent (116) in urdbean and up to 15% in mungbean (33). Narayanaswamy and Jaganathan (153) reported 40% seed transmission of BLCV in blackgram. The extent of seed transmission is influenced by several factors. It is negatively correlated with the age of the plant at the time of inoculation. No transmission is observed if the plants are inoculated at the age of 50 days (154, 155, 211). In case of inoculations done at the age of 5,30 and more than 30 days, the per cent seed transmission is reported 41.6, 20.3 and 3.6% respectively. The seed transmission is not correlated with the morphological abnormalities produced in seed due to the infection,different maturation stages of seed, presence or absence of seed coat, location of pod in the plant or the location of seed in the pod (34, 211). However, the extent of seed transmission may differ from variety to variety (33) of greengram.

The virus has been detected in plumule and radicle of black-gram after 72 hours of soaking the seed in the water and from plumule, radicle and cotyledons after 96 hours of soaking. The presence of the virus is not detected in cotyledons alone indicating that the virus is carried internally in the seed. It is not detected in seed coat also (153, 211). On the other hand Beniwal and Chaubey (32) could detect the virus in primary axis, cotyledons as well as seed coat of blackgram.

The presence of ULCV in seed sample is well represented by presence of oversized seeds in the test samples of seed. It can be further confirmed by `growing on' and indicator inoculation tests

(31). Methods have been described for artificially inoculating the seed with ULCV for use in the research work. Amin *et al.* (13) inoculated urdbean seeds by vacuum infusion of extract in to the water soaked seeds and reported 40% seedlings showing symptoms after 6 weeks. Chowdhury and Nath (61) described a method to inoculate the germinating seeds by shaking them with preparation of ULCV. Highest infection was obtained (90%) when 48 hours germinating seed was shaken with the inoculum for 30 seconds.

(iv) Aphid transmission : *Aphis craccivora* Koch. was first identified as the vector of ULCV (45, 112). Other important species of aphids reported to be the vector of ULCV are *Aphis gossypii* (75), *Acyrthosiphon pisum* Harris (41) and *Myzus persicae* (76). However, none of them could transmit the ULCV at Pantnagar (27, 114, 116).

Aphis craccivora seems to be the better vector than *A. gossypii* (75). Higher percentage of transmission of ULCV is obtained by allowing it to feed on diseased, malformed bunch instead of leaf for a period of few seconds only. Maximum transmission is obtained by allowing 4 hours pre-acquisition fasting and 30 seconds of acquisition feeding (76). Transmission decreases with increasing the acquisition periods and reaches zero per cent when 10 minutes acquisition feeding is permitted. After acquisition, the aphid can transmit the virus immediately but looses its viruliferous nature with in 10-30 minutes. Ten aphids per plant are required for maximum transmission. Adult apterous aphids are more efficient vector than alate adults and nymphs. For transfering the aphid, the tapping technique is superior over the brush method (76).

In case of *Acyrthosiphon pisum*, maximum transmission of ULCV has been observed with a pre-acquisition fasting of 90 minutes (41, 42, 77). Though, the aphid can acquire the virus within a period of 1 minute but for maximum transmission 10 minute acquisition feeding is required. Longer access period leads to decrease in transmission possibly due to the inhibitory saliva. A post acquisition fasting period of 80 minutes is optimum and their viruliferous nature is lost in 60 minutes if they are fed on non host plant like maize, but it is 320 minutes if they are kept under fast (42). The temperature, nutritional status of the plant and the inoculation feeding period also influence the degree of transmission, *A. craccivora* as well as *A. pisum* transmit the virus in probes lasting 1 minute, the optimum being 1-5 minutes and 10-20 minutes respectively. The transmission of ULCV is best if both the aphid species are mixed in a ratio of 1:4 (42).

Accoridng to Bhardwaj and Dubey (42) the relation between *A. craccivora* and ULCV is different than that observed previously (75, 76). They mentioned the optimum acquisition period : 20 minutes, optimum post acquisition fasting : 20 minutes and period for loosing the viruliferous nature : 30 minutes which do not agree with the results obtained previously by Dhingra and Chenulu (76).

(v) Beetle transmission : The beetle, *Epilachna dodecastigma* has been reported to transmit the ULCV in urdbean (27). One beetle allowed per test plant is capable of transmitting the virus with 60 % or more transmission efficiency but 5 or more are required for 100 per cent transmission (44). The optimum period of feeding for acquisition and inoculation is 24 and 48 hours respectively. Pre-acquisition fasting of the adult beetles does not influence the transmission efficiency. Grubs can also transmit the virus but they are less efficient than adults. The beetles are able to retain the virus for 9 days. The virus is not detected in mouth parts, legs and eggs but is readily detected in the faeces of the viruliferous beetles.

(vi) White fly transmission : Transmission of BLCV has been reported in blackgram by the white fly, *Bemisia tabaci* (149). Generally the viruses transmitted by sap are not transmitted by *B. tabaci* but BLCV (ULCV ?) seems to be the exception. For better transmission, the white flies need 1 hour pre-acquisition fasting, 6 hours acquisition feeding and 16 hours inoculation feeding. No incubation period is required in the insect and the transmission is successful by using 15-20 white flies per test plant. The symptoms appear 20-23 days after inoculation feeding.

Variability in the pathogen : The detailed studies on the properties, host range and the transmission show that the isolate of Urdbean Leaf Crinkle Virus used at Pantnagar by Dr. Y.L. Nene and his associates and that used in Tamil Nadu By Dr. P. Narayanaswamy and T. Jaganathan are different from each other. The Pantnagar isolate did not infect *Cajanus cajan, Cyamopsis tetragonoloba* and *Arachis hypogaea* but they were infected by the isolate from Tamil Nadu. Similarly the Pantnagar isolate was not transmitted by white fly and aphids but Tamil Nadu isoalte was transmitted by white flies which is an exception in case of sap transmissible virus like ULCV. It shows that these two isolates are very much different from each other. At Delhi also the isolate of ULCV was transmitted by species of aphids (75, 76) which were not successful at Pantnagar. It indicates the existence of great variability among the isolates of Urdbean Leaf Crinkle Virus.

Host pathogen interaction :

The infection of host plant by ULCV results in morphological, physiological and biochemical changes in the host and disturbs the functioning of certain organales in addition to the visible outer symptoms produced on the plant.

Histological changes : Infection of the urdbean plant by ULCV causes degeneration of the androecium and gynoecium, incomplete development of pollen grains in pollen tetrads, non formation of ovules and abnormal ovarian cavity (212, 213). However, differentiation of cells for pollen mother cells and ovary formation occurs in flower buds if plant is infected late. In leaves, the disease causes reduction in palisade parenchyma, repeated cell division at the margin of the epidermis, increase in number of xylem vessels with more thickened walls and reduction in phloem. The size of the xylem vessels is reduced, cortical and phloem cells are disintigrated and the cell wall of most of the tissues is thickened. It is also reported (39) that cells infected by ULCV are enlarged, the plasmalemma invaginate at certain places giving rise to the formation of vesicles and the mitochondria appear filiform as compared to spherical in healthy cells.

The virus like particles (VLP) are observed in the nucleus, cytoplasm and chloroplasts of ULCV affected leaf cells after 14 days of infection (39). The VLP are spherical, 25-30 nm in diameter and more denser and bigger in size than ribosomes of the cell. They are not seen in the healthy plant or in leaves after 6 or 25 days of infection. The cell sap has the maximum infectivity during the period of appearance of VLP which is 12 days after infection.

Biochemical changes : The infection of urdbean by ULCV increases phosphorus and potassium but decreases nitrogen, calcium and magnesium contents of leaves (269). It reduces RNA and DNA contents of greengram and blackgram leaves but reduction in the protein, carotenoids, chlorophyll `a' and `b' is not much (240). It also reduces the dry weight of plant, O.D. phenols, ascorbic acid and the protease while total phenols, reducing sugars, non reducing sugars are increased (215). The polyphenol oxidase activity is reduced at the initial and severely crinkled stage of the disease. The diseased plant synthesizes glycine in addition to the other amino acids synthesized by the healthy plant. The concentration of other amino acids is also affected in the diseased tissues. The infection causes reduction in tryptophane and increase in IAA contents of leaves which ultimately results in the expansion and crinkling of leaves expressing the main symptoms of the disease (38).

Changes in pollens : Sterility of flowers and reduced pod setting is one of the important symptoms of leaf crinkle disease. It is reported (116, 118) that 30 % flowers on the diseased plants have complete pollen sterility while 70 % buds contain about 39.2 % sterile pollens as compared to 3.6 % pollen sterility in healthy plants. The pollen sterility due to BLCV may differ from variety to variety. In CO 1 and Co 2 varieties of blackgram, it has been observed 71.6 and 13.6 % respectively (153). Similarly early infected plants have high percentage of sterile pollens as compared to late infected ones of the same variety.

Disease cycle and epidemiology :

Urdbean leaf crinkle affected plants appear scattered in the field indicating that the disease develops primarily by the seed borne virus (35). The virus may survive on weed hosts also. For example, *Convolvulus arvensis* is infected systemically by ULCV and its prevalence in a field causes more incidence of ULCV on urdbean as compared to the field free from it (105). The secondary spread of disease may take place by rubbing of the diseased and healthy leaves together particularly late in the season, though it may not cause fast spread of the disease as noted at Pantnagar (116). However, at some of the places the disease is becoming serious as in Haryana and Punjab (22). Similarly in Tamil Nadu, BLCV has been reported to spread very fast which seems to be due to its vector, white fly, detected there (149). In Haryana, India, the urdbean leaf crinkle disease is more prevalent in northern parts than in the southern parts of the state. Also the disease is more severe on rainy season crops as compared to summer season crops of blackgram and greengram (104).

During summer the disease either does not appear or the symptoms are very mild (301). The maximum disease is observed at conditions with maximum temperature 35°C (±2°C), minimum temperature 25°C (±2°C), minimum RH more than 70 % and evening RH more than 50 %. In summer the symptoms are masked at 38-45°C when morning and evening RH remain 60 and 40 % respectively. The disease symptoms do not appear even at 35° C if host plants are already exposed to 41-45° C for a week or to more than 47°C for a day with morning and evening RH of 45 and 20 % respectively.

Host resistance :

Large number of germplasm and cultivars of greengram, black gram and cowpea possess resistance against ULCV (Table 4.3). In general, greengram is comparatively more resistant than blackgram.

Table 4.3

Cultivars and germplasm types of different crops showing resistance to ULCV/BLCV

Crop	Entries tested	Resistant entries	Reference
Blackgram	-	N 212, 1766	(14)
	52	No. 212 with less disease	(161)
	68	Karaikal, Polladam, Mattikalai, Parvatipuram, BR 16, BR 68, were free and Kahandur, PHM 8, P 33, P 49, P 58, P 223, P 27 and P 38 resistant	(150)
	288	HPU 27, HPU 102, HPU 164 and HPU 35 were resistant	(214)
Greengram	12	24-3, T 44, T 2, T 51 and Pusa Baisakhi were free	(161)
	-	ML 26, ML 9, 11157 and ML 70-10	(101)
	39	D 215, HPM 1, Madana-1, MS 8, ML 12, Progeny 5 were immune; Bhabhar, D 3-9, J 1, K 12, ML 26, S 9 and T 44 were resistant	(214)
	-	V 2182 and V 2294 (from AVRDC)	(237)
Cowpea	8	JC 10, Mashad, No. 1-1, No. 5286-3, PPSC 1 and PPSC 3 were free	(161)

For screening purposes the disease scoring may be done on the basis of plants infected at seedling and pod formation stages as per scale given below :

Scale grade	Particulars of disease
0	No plant showing any symptom
1	Less than 1 % plants showing symptoms
3	1-10% plants showing symptoms
5	11-20 % plants showing symptoms of crinkling
7	21-50% plants showing crinkling of leaves and stunting of plants.
9	50% or more plnats are affected with severe leaf crinkling, stunting of plants and malformation of inflorescence.

Control :

Source of seed : Since, the ULCV is seed transmitted, the seed should not be procured from the diseased crop. The disease free seed can be grown by controlling the disease or atleast by delaying the infection to a longer possible period because the extent of seed transmission depends on the stage of the crop at which plants are infected.

Seed treatment : The seed treatment by wet heat (hot water) is found best for inactivating the seed borne ULCV without adversely affecting the seed germination. For this purpose temperature level of 55°C for 30 minutes (214), 60°C for 10 to 30 minutes or 70°C for 10 to 20 minutes (34), 65°C for 30 minutes or 70°C for 20 minutes (102) have been reported to be effective. Soaking seed in water fo 3-4 hours and then exposure to solar heat from 12 to 4 P.M. in May and June is also effective in eliminating the seed borne infection of ULCV (102). Dry heat treatment at 70°C for 10 to 30 minutes is effective in cheking the transmission of the disease through seed (34) but it affects the germination of seed adversely (102).

Chemotherapy : The seed dip in 2-thiouracil or 8-azaquinine 0.05 and 0.1% concentration for 3 minutes has been found effective (102). Drenching of greengram plants with 1 % benomyl before their inoculation, mechanically or by viruliferous aphids, prevent the development of symptoms (43). Aphids are unable to acquire the ULCV from plants drenched with 2 % benomyl or bavistin but drenching with vitavax was phytotoxic (214). Pre- and post-inoculation spray with 200 ppm thiouracil delays the appearance of symptoms while IAA enhances the disease. The actinomycin D and cycloheximide decrease the infectivity of ULCV when sprayed on to the plant immediately after the inoculation.

Use of plant extracts : Decoction of tea and coffee, used as soil drench, is phytotoxic to urdbean plants (214). On the other hand extracts of zinger and turmeric inhibit the ULCV *in vitro* to an extent of more than 50 per cent (62). Pre-inoculation spray of onion extract showed maximum inhibition of ULCV *in vivo*.

Vector control : Aphids, white flies as well as beetles are reported vectors of ULCV. Though, no experiments have so far been conducted, chemical control of vectors particularly during early stages of crop growth is expected to minimize the losses caused by the Urdbean Leaf Crinkle Virus.

Cultural : Since, the disease is easily transmitted mechanically, destruction of diseased plants at early stage and avoiding contacts between diseased and healthy plants, particularly rubbing of leaves together during intercultural operations, will minimize the disease spread in the field. Proper weeding and elimination of weed hosts from the area is also essential for controlling the disease.

4.4. LEAF CURL OF MUNG AND URD :

The occurrence of leaf curl disease of urdbean was reported as early as 1921 in the Dutch East Indies and the damage caused by the disease was noted but the cause of the disease was not described in detail (285). In India, the leaf curl of mungbean and urdbean was observed and studied in detail (159, 161). On the basis of its symptomatology, the disease was suspected of the viral origin (159). A leaf curl browning disease of mungbean was reported from Philippines with more or less similar symptoms (25). A leaf curl disease distinct from BCMV and Yellow bean mosaic or curly top was observed on mungbean, urdbean and french bean (258). In India, the leaf curl disease has been reported from Uttar Pradesh (161), Andhra Pradesh (16) and Delhi (286).

Nature and extent of damage :

Among the viral diseases of mungbean and urdbean, leaf curl is potential killer of plants. Majority of plants showing symptoms within 5 weeks of sowing die due to top necrosis. Plants which escape the death, remain stunted and fail to flower. Even late infection reduces number of pods which either produce smaller and lighter grains or no grains at all. Grains produced by diseased plants are unfit for seed and consumption purposes. Thus, every plant affected by the disease at any stage of crop growth becomes economically useless. The extent of such affected plants may range from 10 to 42 per cent (161).

Symptoms :

Symptoms of the disease appear throughout the season starting within 20 days of sowing till the plants are able to throw last new leaves (161). First symptoms appear in the form of chlorosis around some lateral veins and its branches near the margin of youngest leaves. Margin of leaves curl downward and rolling and twisting of leaves is seen in some cases. The leaves become brittle and the whole trifoliate including the stalk may fall down from the plant on giving a jerk. Leaves of the diseased trifoliate are far removed from each other. The veins show reddish brown discolouration on the under surface which also extends to the petiole. Affected plants stop further growth and can easily be spotted out due to their typical stunted growth in early stage of the crop growth (Fig. 4.4). Plants showing symptoms within 5 weeks after sowing remain stunted and mostly die due to top necrosis within a week or two. Plants which escape death, develop conspicuously smaller leaves and fail to flower. In case of late infections, severe curling and twisting of leaves is not produced but veinal chlorosis becomes prominent any where on the lamina. Such plants can produce some pods. Leaves showing initial symptoms without distortion, show ringspot symptoms late in the season. Rings which are initially chlorotic, turn necrotic later. The leaf curl may be found simultaneously with yellow mosaic, leaf crinkle or root knot and the symptoms may be different on different varieties. In short, the characteristic symptoms of the leaf curl are down ward curling of leaves, veinal necrosis, chlorotic areas on the lamina, stunting and death of plant due to apical necrosis (83).

Fig.4.4.
Leaf curl of black gram.

/irus :

Tomato Spotted Wilt Virus (TSWV) Ghanekar *et al.* 1979.

In the begining, Nene (159) suggested the disease to be of viral origin but he did not identify the causal virus. Later, on the basis of detailed study done on the serology, transmission, host range and other properties of the virus it was identified as Tomato Spotted Wilt Virus (83). The thermal inactivation point of the virus is 45-50°C and the dilution end point is between 10^{-2} to 10^{-3}. The longevity *in vitro* of the crude infected sap is 5 h at room temperature i.e. 25°C (16). The causal virus of the leaf curl of urdbean showed strong serological affinity with Tomato Spotted Wilt Virus. The hemagglutination titers of leaf curl and TSWV is 1/8000. In the ELISA test, the leaf curl and TSWV had a titre of 1/400 and 1/800 respectively with the antiserum to TSWV isolate causing bud necrosis disease of groundnut.

Transmission : The causal agent has been shown to be transmitted by grafting from mung bean to mungbean and urdbean producing symptoms within 10-15 days after grafting (161).

(i) **Mechanical transmission:** Inoculation of [illegible] the help of carborundum produces local lesions [illegible] are initially chlorotic but later turn necro[illegible] chlorosis and curling of young, non [illegible] lesion symptoms (161).

(ii) **Vector transmission** : The thrip, *Frankliniella schultzii* (Trybon.) is reported to transfer the TSWV of groundnut as well as leaf curl of mung and urdbean (16). Transmission rates are high if the test plants are mungbean or urdbean as compared to groundnut. Source of the virus does not affect the transmission rate much. The incubation period is 5-6 days in mung and urd beans as compared to 12-15 days in groundnut. The acquisition access, latent and inoculation access periods are 24 h, 5-7 days and 5 days respectively. The virus is not transmitted by the whitefly, seeds and the soil. Though, the early appearance of the disease and the irregular distribution of diseased plants in the field are the factors to suspect the virus to be seed borne but it has not been confirmed (161).

Host range : Cowpea variety C 20 has been reported as the local lesion host of TSWV showing chlorotic as well as necrotic lesions after 4-5 days of inoculation (82). Later, the virus becomes systemic causing tiny necrotic lesions, downward curling, veinal necrosis, reduction in leaf size and top necrosis of the plant. Out of 23 plant species tested, *Vigna mungo* (cv. UPU 1), *V. radiata, V. unguiculata* (cv. Top crop), *Nicotiana clevelandii* and *Datura stramonium* showed all the three types of symptoms (chlorotic local lesions, necrotic local lesions and systemic infection) when inoculated by TSWV. On *Vicia faba, Nicotiana glutinosa* X *N. clevelandii, Lycopersicon esculentum* (cv. Pusa ruby) and *Canavalia ensiformis*, the virus produced necrotic local lesions and systemic infection. In case of *Arachis hypogaea* (cv. TMV 2), *Glycine max* (cv. Bragg) chlorotic local lesions and systemic infection were recorded while on *Cajanus cajan* (ICP 2376), *Chenopodium amaranticolor, Chenopodium quinoa, Gomphrena globosa, Nicotiana tabacum* (cv. Xanthinc), *Petunia hybrida* (cv Coral satin) and *Solanum tuberosum*, only necrotic local lesions were produced. On *Nicotiana rustica*, chlorotic and necrotic local lesions were formed but *Penisetum americanum* (cv. BK 560) and *Sorghum vulgare* (cv. CSH 6) were not infected (16).

Host resistance :

Large number of greengram and blackgram lines collected from various localities and several cultivars of both the crops were screened against the leaf curl virus to find the sources of resistance (161). Some of them showed resistance as indicated below :

Mungbean : Hy 45, ST 7, P 15, 56, 101, 435-1, 435-2, 435-3, 382, 386, 135-2, Barabanki, Jhansi 2, Jhansi 3, Sitapur line.

Blackgram : N 212, Khargone 3, Lines collected from Jhansi, Lucknow 1, Lucknow 2, Lucknow 4, Pratapgarh 2, Pratapgarh 1, Pratapgarh 3, Shahjahanpur 2, Shahjahanpur 3 and Sitapur.

It is indicated that the resistance exists in both the crops against leaf curl virus. It is desirable to screen large number of collections under artificially inoculated conditions and to identify good sources of resistance for use in resistance breeding programmes.

Control :

No specific control measures are recommended/investigated against leaf curl of greengram and blackgram but control of vector (thrips) by use of insecticide may help in minimizing the disease.

4.5. MUNGBEAN LITTLE LEAF :

The little leaf disease of mungbean was observed in Philippines (25). The virus is not transmitted mechanically but by seed and the vector *A. craccivora* in non persistant manner. The details of the virus are not known but it seems similar or the same as that causing little leaf in cowpea. It may infect the mungbean plant alone or with any or all of the yellow mosaic (Philippine strain), mungbean green mosaic and leaf curl browning. However, all of them can be isolated in to pure culture using differential hosts. Yellow mosaic produces local lesions on mungbean and cowpea when inoculated mechanically. Mungbean green mosaic produces vein necrosis on cotyledonary leaves of mungbean and numerous, faint, small, chlorotic local lesions on *C. amaranticolor.* Leaf curl browning forms large chlorotic local lesions with pin head size, straw coloured, necrotic centres on *C. amaranticolor*. Little leaf virus can be isolated through *A.craccivora* transmission to mungbean in which trifoliate becomes dimunitive in size (25).

4.6. WHITE FLY TRANSMITTED VIRUS DISEASE :

A virus causing severe leaf crinkling and rugosity, bunching of inflorescence and reduction in pod setting was noticed on urd bean in Andhra Pradesh, India (302). The electron microscopic studies and attempts to purify the virus were not successful and serologically it was not found related with at least 11 viruses. Out of 59 host plant species tested, urdbean, mungbean and frenchbean were infected by sap inoculation. In addition, the disease was transmitted by seed (7-20 %) and by white fly in non persistant manner (0-20 %) but not by *A. craccivora*.

4.7. TABACCO NECROSIS VIRUS (TNV) :

The washings of roots of crops infected with *Olpidium brassicae* and TNV were able to cause lesions on mungbean roots in 2 days (275). The letuce strain of *O. brassicae* was able to transmit the virus in mungbean seedling and the amount of infection was proportional to number of zoospores (272,274). The sensitivity of the infection is increased if rub inoculation is done (273). The infective RNA of TNV from frenchbean could be inoculated in hypocotyles of mungbean grown in dark (131) and the virus was inhibited by cycloheximide due to inhibition of the protein synthesis required for the virus multiplication.

4.8. OTHER VIRUSES INFECTING GREENGRAM AND BLACKGRAM :

Several viruses have been found to infect mungbean and urdbean crops during the host range studies and few under natural conditions. They are summarized in Table 4.4.

Table 4.4
Viruses infecting mungbean urdbean crops

Virus/disease	Symptoms on crop	Country/ Institute	Reference
Blackgram sterility mosaic	Reduces blackgram yield	India	(150, 151)
Broadbean Mosaic Virus	Infects mung and urdbean	India	(280, 281)
Broadbean Mottle Virus-blackgram strain	Detected in urdbean seed	India	(176)
Citrus viruses	Local lesions on mungbean	BARI	(127)
Chickpea Distortion Msaic	Mungbean symptom less carrier	India	(128)
Clover mosaic	Sap transmission to mung	New Zea-land	(80)
Coconut Wilt Virus	Wilt in urdbean	India	(210)
Cowpea severe mosaic	Serotype detected in mung	Brazil	(125)
Double bean yellow mosaic	Infected mungbean	India	(48)
Guar top necrosis	Sap transmission to mung		(66)
Mosaic of *Crotalaria mucronata*	Sap transmission to both	India	(198)
Mungbean Mottle Virus	Infected mungbean	Thailand Taiwan	(148) (299)
Mungbean Yellow Mottle Virus	Infected mungbean	Taiwan	(299)
Okra Mosaic Virus	Soil transmission in mung	Nigeria	(123)
Pea Enation Mosaic Virus	Infected mungbean		(47)
Pigeonpea Yellow Mosaic	Urdbean mosaic	Ceylon	(168)
Satellite Tobacco Necrosis	Multiplied in mungbean	-	(126)
Sugarcane Mosaic Virus	Local lesion in mung bean	India	(79)
Tobacco Rattle Virus	Local lesion in mungbean	-	(129)
Tobacco Sreake Virus/Red node disease	Necrosis of urd and mung	Maryland	(276)
Tobacco Ring Spot Virus	Systemic infection and death of mungbean	Indonesia India	(99) (217)
Virus from *Cassia occidentalis*	Mosaic in urdbean	India	(229)
Cowpea (yellow) mosaic	Infects mung and urd	-	(173)
Mosaic of mungbean	Detected on mung bean seed in plant quarantine lab.	india	(184)

4.9. PHYLLODY :

The phyllody diseases of crops are chracterized by transformation of floral parts in to leafy structures. However, it does not represent the symptom complex perfectly. Hence, Bos suggested the term `Antholysis' in place of Phyllody which includes four types of phenomenon. `Virescence' (greening of all the floral parts) `Phyllody' (transformation of segments in to leafy structures), `Apostasis' (elongation of torus internodes) and `Proliferation' (development of leafy branches in the axils of the floral parts). All these symptoms have been observed on greengram in Khartoun, Sudan (169). The typical phyllody disease of mungbean was first observed in Uttar Pradesh, India in the year 1952 (232). The disease is potentially dangerous, though its incidence in north India is reported to an extent of 0.5 % only (55). The disease has been reported occurring on blackgram also (221).

Symptom :

All the flower parts of mungbean are transformed in to leaf like structures (232). The petals turn green, veins become marked, petals are very much rounded and the standard, wings and keel become separated exposing the ovary. Flowers assume the erect position with elongated ovary which fails to develop the ovule. In some flowers a shoot like the main inflorescence arises in the axil of the ovary while in others a central shoot arises from the middle of the flower bearing phylloid flowers at the top. The stemans are aborted and anthers do not dehisce. In some flowers, stemans probably turn in to leaf like structures and inflorescence on the whole gives a rossette type of appearance.

More or less similar symptoms are produced on blackgram (221). In severe cases, all the flower buds are affected and the plant remain stunted and unbranched. On some plants, a few yellow flower buds may be seen but they do not open normally or produce the abnormal seedless pods of small size at maturity.

Pathogen :

Mycoplasma Like Organism (MLO).

In Sudan, the Phyllody of greengram showing green flowers and phyllody symptoms with smalling and yellowing of leaves was considered to be due to a virus belonging to yellows group which is not transmitted by sap (169). Though, no detailed studies have been made but now the disease is known to be caused by Mycoplasma like organism (179).

Control measures and other details are not known.

4.10. WITCHES BROOM :

It is the minor disease of greengram observed first at AVRDC Taiwan in the field trials (299), though, reference of a systemic brooming disease (witches broom) of blackgram occurring in Java has been made as early as 1913 (92). The disease is reported to occur in Philippines where it is caused by the Mycoplasma Like Organism (MLO) transmitted by leaf hoppers (23).

The symptoms of the disease include excessive branching, dwarfed leaflets and malformed flowers with excessive proliferation of axillary buds. Infected plants are usually sterile.

REFERENCES

1. Agrawal, H.S., Gupta M.K., Prasad V.K. and Vishwakarma S.L. 1979. Chemical control of yellow mosaic of moong. *Pesticides*, 13 (5) : 44-47.

2. Agrawal, S.C. and Nema S. 1984. Plant Pathology. *Annual Report, 1983-84, All India Coordinated Pulses Improvement Project, RAK Agriculture College Sehore, India,*30-41 pp.

3. Agrawal, V.K., Beniwal S.P.S. and Verma H.S. 1976. Note on transmission of bean common mosaic virus in seeds of green gram (*Vigna radiata* (L.) Wilczek). *Pantnagar Journal of Research,* 1: 81-82.

4. Agrawal, V.K., Nene Y.L. and Beniwal S.P.S. 1976. Influence of bean common mosaic virus infection on the flower organelles, seed characters and yield of urid bean. *Indian Phytopathology*, 29 : 444-445.

5. Agrawal, V.K. Nene Y.L. and Beniwal S.P.S. 1977. Detection of bean common mosaic virus in urid bean(*Phaseolus mungo*) seeds. *Seed Science and Technology* 5 : 619-625.

6. Agrawal, V.K., Nene Y.L. and Beniwal S.P.S. 1979. Location of bean common mosaic virus in the urid bean seed . *Seed Science and Technology*, 7 : 455-458.

7. Agrawal, V.K., Nene Y.L. Beniwal S.P.S. and Verma H.S. 1979. Transmission of bean common mosaic virus through urd bean seeds. *Seed Science and Technoloy*, 7: 103-108.

8. Ahmad, M. 1975. Screening of mung bean (*Vigna radiata*) and urid bean (*Vigna mungo*) germplasm for resistance to yellow mosaic virus. *Journal of Agricultural Research Pakistan,* 13 : 349-354.

9. Ahmad, M. and Harwood R.F. 1973. studies on a white fly transmitted yellow mosaic of urid bean (*Phaseolus mungo*). *Plant Disease Reporter,* 57 : 800-802.

10. Ahmad, S.R. and Gaur R.B. 1982. Chemotherapuetic approach to check yellow mosaic virus of moong (*Phaseolus aureus* L.). *Hindustan Antibiotics Bulletin,* 24 (1/2) : 18-20.

11. Ahmed, H.U., Bakr M.A., Alam K.B. and Kaur A.K. 1981. Diseases of Major pulse crops in Bangladesh. *III International Symposium of Plant Pathology*, 86 pp.

12. Ahuja, M.R. and Singh B.V. 1977. Induced genetic variablitity in mung bean through interspecific hybridization. *Indian Journal of Genetics and Plant Breeding,* 37 : 133-136.

13. Amin, K.S., Baldev B. and Williams F.J. 1978. Transmission of leaf crinkle virus of urid bean by vacuum infusion of extract in to water soaked seeds. *Current Research,* 6: 98-99.

14. Amin, K.S., Grewal J.S., Verma A. and Williams F.J. 1968. Screening of mung and urid genetic stock for disease resistance. *Proceedings II Workshop Conference on Pulse Crops, IARI, New Delhi.*

15. Amin, K.S. and Singh R.A. 1987. Diseases of mung urid and pea and their control. *In Plant Protection in Field Crops (ed. M. Veer Bhadra Rao and S. Sithananthan), Plant Protection Association of India,* 203-217 pp.

16. Amin, P.W., Ghanekar A.M., Rajeswari R. and Reddy D.V.R. 1985. Tomato spotted wilt virus as the causal pathogen of leaf curl of mung bean, *Vigna radiata* (L.) Wilczek and urd bean, *Vigna mungo* (L) Hepper in Andhra Pradesh, India. *Indian Journal of Plant Protection,* 13 : 9-13.

17. Anonymous. 1945. *Report of Imperial Council of Agriculture Research, Delhi 1944-45.* 1-11 pp.

18. Anonymous, 1952. Report of the Division of Mycology and Plant Pathology. *Scientific Report Agriculture Research Institute New Delhi,* 1949 - 50, 81-82 pp.

19. Anonymous. 1962. Report of Indian Agricultural Research nstitute, New Delhi, 1962, 2 : 147-159.

20. Atheya, S.C. 1977. A mosaic disease of urid from Kanpur, India. *Indian Journal of Mycology and Plant Pathology* 7 : 99-100.

21. Babu, R.C., Rathinaswamy R., Srinivasan P.S., Natrajaratnam N. and Sreerangaswamy S.R. 1984. Certain Physiological changes in green gram plants infected by mungbean yellow mosaic virus. *Madras Agricultural Journal* 71 : 795-798.

22. Bansal, R.T., Khatri H.L., Sharma O.P. and Singh I. P. 1984. Epidemiological studies on virus diseases of mung and mash in Punjab. *Journal of Research Punjab Agricultural University* 21 : 54-58.

23. Benigno, D.R.A. 1979. Occurrence of witches broom disease of mung bean in Philippines. *Philippine Phytopathology.* 15 : 86.

24 Benigno, D.R.A. 1979. Note : an aphid -borne virus isolated from black gram in the Philippines. *Philippine Agriculturist,* 62 : 328 - 332.

25. Benigno, D.R.A. and Dolores A.C. 1977. Virus diseases of mung bean in the Philippines. *I International Mungbean Symposium, AVRDC, Taiwan,* 173-175 pp.

26. Benigno, D.R.A. and Favali Hedayat M.A. 1977. Investigations on previously unreported or noteworthy plant viruses and virus diseases in the Philippines. *FAO Plant Protection Bulletin,* 25(2) : 78-84.

27. Beniwal, S.P.S and Bharathan N. 1980. Beetle transmission of urd bean leaf crinkle virus.*Indian Phytopathology,*33 : 600-601.

28. Beniwal, S.P.S. Bharathan N. and Chaubey S.N. 1983. Two cucurbitous hosts of urd bean leaf crinkle virus. *Indian Phytopathology,* 36, : 577-579.

29. Beniwal, S.P.S and Chaubey S.N. 1979. Transmission of urd bean strain of bean common mosaic virus (UB BCMV) by a tarsonemid mite, *Polyphagotarsonemus latus. Indian Phytopathology,* 32, : 163.

30. Beniwal, S.P.S. and Chaubey S.N. 1979. Urd bean Leaf crinkle virus : Effect on yield contributing factors, total yield and seed characters of urd. *Seed Research,* 7 : 175-181.

31. Beniwal, S.P.S. and Chaubey S.N. 1984. Detection of urd bean leaf crinkle virus in urd bean seed. *Seed Research,* 12 : 101-104.

32. Beniwal, S.P.S and Chaubey S.N. 1984. Internal seed borne nature of urd bean leaf crinkle virus in urd bean seed. *Seed Research* , 12 : 8-10.

33. Beniwal, S.P.S. Chaubey S.N. and Bharathan N. 1980. Presence of urd bean leaf crinkle virus in seeds of mung bean germplasm. *Indian Phytopathology,* 33 : 360-361.

34. Beniwal, S.P.S., Chaubey S.N. and Matheswaran C. 1983. Some factors affecting transmission of urd bean leaf crinkle virus through seeds of urd bean (*Vigna mungo (L.) Hepper). Seed Research,* 11 : 95-99.

35. Beniwal, S.P.S., Kolte S.J. and Nene Y.L. 1979. Nature and rate of spread of urd bean leaf crinkle disease under field conditions. *Indian Journal of Mycology and Plant Pathology,* 9 : 188-192.

36. Beniwal, S.P.S., Vohra K. and Verma H.C. 1980. Effect of mung bean yellow mosaic virus infection at different stages of plant growth on the amino acid composition of urd bean *(Vigna mungo L.)* seeds. *Indian Journal of Experimental Biology,* 18 : 302-303.

37. Bhaktavatsalam, G. and Beniwal S.P.S. 1977. Note on auxin content in urd bean (*Phaseolus mungo L.*) infected with mung bean yellow mosaic virus. *Pantnagar Journal of Research,* 2 (2) : 234-235.

38. Bhaktavatsalam, G., Nene Y.L. and Beniwal S.P.S. 1982. Hyper auxiny in urd bean eaf infected by urd bean leaf crinkle virus. *Indian Phytopathology,* 35 : 683-685.

39. Bhaktavatsalam, G., Nene Y.L. and Beniwal S.P.S. 1983. Ultra structural changes in urd bean leaves infected by urd bean leaf crinkle virus. *Indian Phytopathology,* 36 : 228-232.

40. Bhaktavatsalam, G., Nene Y.L. and Beniwal S.P.S. 1983. Influence of certain physico-chemical factors on the infectivity and stability of urd bean leaf crinkle virus. *Indian Phytopathology,* 36:489-493.

41. Bhardwaj, S.V. 1981. Some factors affecting transmission of urd bean *(Vigna radiata* var. *mungo)* leaf crinkle virus by aphids. *III International Symposium of Plant Pathology IPS, New Delhi,* 165 pp.

42. Bhardwaj, S.V. and Dubey G.S. 1986. Studies on the relationship of urd bean leaf crinkle virus and its vectors, *Aphis craccivora and Acyrthosiphon pisum. Journal of Phytopathology,* 115 : 83-88.

43. Bhardwaj, S.V., Dubey G.S. and Sharma I. 1982. Effect of benlate on infection and transmission of urd bean *(Vigna radiata* var. *mungo)* leaf crinkle virus. *Phytopathologische Zeitschrift,* 105 : 87-91.

44. Bharathan, N. and Beniwal S.P.S. 1984. Transmission characteristics of urd bean leaf crinkle virus by Epilachna beetle, *Henosepilachna dodecastigma. Indian Phytopathology,* 37 : 660-664.

45. Bindra, O.S. 1971. Studies on Arthropodes in relation to plant diseases in Punjab. *II International Symposium on Plant Pathology, IARI, New Delhi.*

46. Bisht, I.S., Singh Shamsher and Dabas B.S. 1988. Effect of yellow mosaic virus on growth components and yields of four promising cultures of mung bean. *Indian Journal of Mycology and Plant Pathology,* 18 : 204-205.

47. Bustrillos, A.D. 1965. Purification, serology and electron microscopy of pea enation mosaic virus. *Diss. Abstract,* 25 (11) : 6145-6146.

48. Capoor, S.P. and Varma P.M. 1984. Yellow mosaic of *Phaseolus lunatus* L. *Current Science,* 17 : 152-153.

49. Chand, P. and Verma J.P. 1980. Some characteristics of mung bean and urd bean varieties resistant and susceptible to yellow mosaic virus. *Indian Phytopathology,* 33 : 48-53.

50. Chand, P. and Verma J.P. 1983. Effect of yellow mosaic on growth components and yield of mung bean and urd bean. *Haryana Agricultural University Journal of Research,* 13 : 98-102.

51. Chandel, K.P.S., Joshi B.S. and Pant K.C. 1980. Breeding mung bean (*Vigna radiata* (L) Wilczek) for photoinsensitivity, high yield and field resistance to Yellow mosaic virus. *Tropical Grain Legume Bulletin* NO. 17/18 : 29-32.

52. Chaudhary, G.G., Mathur A.K. and Tyagi R.N.S. 1981. Reaction of mung cultivars to yellow mosaic virus of mung (*Phaseolus aureus*) in Rajasthan. *Indian Journal of Mycology and Plant Pathology*, 11 : 273-274.

53. Chenulu,V.V. 1984. Plant Virology in Indiia - Past, present and future. *Indian Phytopathology*, 37 : 1-20.

54. Chenulu, V.V., Venkateswarlu V. and Rangaraju R. 1979. Studies on yellow mosaic disease of mung bean. *Indian Phytopathology*, 32 : 230-235.

55. Chenulu, V.V. and Varma Anupam. 1988. Virus and virus like diseases of pulse crops commonly grown in India. In "Pulse Crops" (Edited B. Baldev, S. Ramanujam and H.K. Jain). Oxford & IBH Publishing Company Private Limited, New Delhi, 338-370 pp.

56. Chhabra, K.S. and Kooner B.S. 1980. Sources of white fly and yellow mosaic virus resistance in *Vigna radiata* Wilczek, *Tropical Grain Legume Bulletin* No. 19 : 26-29.

57. Chhabra, K.S., Kooner B.S. and Singh Gurdip. 1979. Field resistance of certain cultivars of mung bean to white fly, *Bemisia tabaci* G. and yellow mosaic virus. *Journal of Research Punjab Agricultural University*, 16: 385-388.

58. Chohan, J.S. and Kaliha N.R. 1967. Virus diseases of *Phaseolus mungo* L. and their control through resistant varieties in Punjab. *All India Pulse Workers Conference, New Delhi*, 65-67 pp.

59. Chowdhury, A.K. and Chowdhury S.R. 1983. Effect of buffers of different pH on the infectivity of the urid leaf crinkle virus (ULCV). *Geobios*, 10 : 34-35.

60. Chowdhury, A.K. and Chowdhury S.R. 1983. Effect of uridbean leaf crinkle virus, ULCV, on plant morphology and yield contributing characters of cultivar TG *Pulse Crop Newsletters*, 3 : 54-55.

61. Chowdhury, A.K. and Nath P.S. 1983. A rapid method of inoculation of urid bean leaf crinkle virus (ULCV) using germinated seeds. *Indian Journal of Experimental Biology*, 21 : 158.

62. Chowdhury, A.K. and Saha N.K. 1985. Inhibition of urdbean leaf crinkle virus by different plant extracts. *Indian Phytopathology*, 38 : 566-568.

63. Chowdhury, J.R. and Srivastava R. S. 1986. Effect of common bean mosaic virus infection on nitrate reductase activity in leaf tissues of mung bean. *Zeitschrift fur Pflanzenkrankheiten und Pflanzenschutz*. 93: 33-36.

64. Chowdhury, J.R., Srivastava R.S. and Singh R. 1985. Effect of common bean mosaic virus and *Rhizobium* on the amino acid contents in mung bean (*Vigna radiata* (L.) Wilczek). *Rivista di Agricoltura Subtropicale e Tropicale,* 79 : 411-424.

65. Chowdhury, J.R., Srivastava, R.S. and Singh R. 1987. Effect of common bean mosaic virus infection on nitrogenase activity in root nodules of mung bean. *Zeitschrift fur Pflanzenkrankheiten und Pflanzenschutz,* 94 : 126-129.

66. Cooper, W.E. 1949. Top necrosis, a virus disease of Guar. *Phytopathology*, 39 : 347-358.

67. Cortalo, R.V. 1972. The control of mungo mosaic by insecticide spray against the aphid. *Philippines Phytopathology,* 8 : 2.

68. Dahiya, B.S., Singh K. and Brar J.S. 1977. Incorporation of resistance to mung bean yellow mosaic virus in black gram (*Vigna mungo* L.). *Tropical Grain Legume Bulletin* No. 9 : 28-32.

69. Dale, W.T. 1943. Preliminary studies of the plant viruses of Trinidad. *Tropical Agriculture Trinidad,* 20 : 228-235.

70. Dale, W.T. 1949. Observations on a virus disease of cowpea in Trinidad. *Annals of Applied Biology*, 36 : 327-333.

71. Das Gupta, B. and Chowdhury A.K. 1985. Use of intervarietal intercropping to minimize the yellow mosaic virus of urd and mung beans. *Indian Journal of Plant Pathology,* 3 : 100-101.

72. Dey, S.K. and Singh D.R. 1973. Resistance test of mung (*Phaeolus aureus* L.) against yellow mosaic in Uttar Pradesh. *Indian Journal of Farm Sciences,* 1 : 113.

73. Dey, S.K. Singh R.A. and Shukla T.N. 1970. Varietal reaction against mung yellow mosaic. *Recent Advances in Crop Production, U.P. Institute of Agricultural Sciences, Kanpur, U.P., India,* 248-249 pp.

74. Dolores, A.C. 1975. Studies of a yellow mosaic disease of mungo (*Phaseolus aureus* Roxb.). B.S. Thesis, UPLB, Laguna, Philippines.

75. Dhingra, K.L. 1975. Transmission of urid bean leaf crinkle virus by two aphid species. *Indian Phytopathology* 28 : 80-82.

76. Dhingra, K.L. and Chenulu, V.V. 1981. Studies on the transmission of urid leaf crinkle and chickpea leaf reduction viruses by *Aphis craccivora* Koch. *Indian Phytopathology*, 34 : 38-42.

77. Dubey, G.S., Sharma I. and Pradhan N.1983. Some properties of urid bean leaf crinkle virus. *Indian Phytopathology*, 36 : 762-64.

78. Dwivedi, S. and Singh D.P. 1985. Inheritence of resistance to yellow mosaic virus in a wide cross of black gram (*Vigna mungo* (L.) Hepper). *Zeitschrift fur Pflanzenzuchtung,* 95 : 281-84.

79. Farrag, S.H. and Kandaswamy T.K. 1979. Soybean and green gram as local lesion assay hosts for sugar cane mosaic virus. *Current Science,* 48 : 865.

80. Fry, P.R. 1959. A clover mosaic virus in New Zealand pastures. *New Zealand Journal of Agriculture Research,* 2 : 971-81.

81. Garg, I.D. and Mandahar C.L. 1977. Physiology of *Phaseolus mungo* leaves infected with bean yellow mosaic virus. *Indian Phytopathology,* 30 : 123-124.

82. Ghanekar, A.M. and Beniwal S.P.S. 1975. Cowpea, a local lesion host for mung bean leaf curl virus. *Indian Phytopathology,* 28 : 527-528.

83. Ghanekar, A.M., Reddy D.V.R. and Amin P.W. 1979. Leaf curl disease of mung and urd beans caused by tomato spotted wilt virus. *Indian Phytopathology,* 32 : 163.

84. Gill, A.S., Verma M.M., Dhaliwal H.S. and Sandhu T.S. 1983. Interspecific transfer of resistance to mung bean yellow mosaic virus from *Vigna mungo* to *Vigna radiata. Current Science*, 52 : 31-33.

85. Gill, K.S., Sandhu T.S., Singh K. and Brar J.S. 1975. Evaluation of mungbean germplasm (*Vigna radiata* (L.) Wilczek). *Crop Improvement,* 2 : 99-104.

86. Godse, D.B., More B.B. and Patil P.L. 1976. Effect of common bean mosaic virus infection on nodulation of mung (*Phaseolus aureus* Roxb.). *Journal of Maharashtra Agricultural Universities*, 1 : 310-311.

87. Grewal, J.S. 1977. Diseases of mung bean in India. *I International Mungbean Symposium, AVRDC*, Taiwan, 165-168 pp.

88. Gupta, P.K. and Singh J. 1983. Effect of systemic granular insecticides on white fly population and yellow mosaic infection in green gram. *Indian Journal of Agricultural Sciences*, 53 : 737-742.

89. Gupta, V.K. 1974. Leaf crinkle virus disease of *Phaseolus mungo* L. in Himachal Pradesh. *Indian Journal of Experimental Biology* 12 : 477-478.

90. Gurha, S.N., Misra D.P. and Kamthan K.P. 1982. Studies on some aspects of yellow mosaic disease of black gram (*Vigna mungo* (L.) Hepper). *Madras Agricultural Journal*, 69 : 435-438.

91. Gurha, S.N., Misra D.P. and Singh R.A. 1984. Yield stimulatory potential of temic and disyston in urd. *Pesticides*, 18 (8): 34-35.

92. Hartley, C. and Haasis F.W. 1929. Brooming disease of black locust (*Robinia pseudocacia*). *Phytopathology*, 19 : 162-166.

93. Honda, Y. 1986. Yellow mosaic virus. In International Symposium on virus diseases of rice and leguminous crops in the tropics. *Tropical agricultural Research Centre Japan Series* No. 19 : 121-128.

94. Honda, Y., Iwaki M., Thongameearkom P., Deema N. and Srithongchai W. 1982. Black gram mottle virus occurring on mung bean and soybean in Thailand. *JARQ* 16 : 72-77.

95. Honda, Y., Iwaki M., Saito Y., Thongameearkom P., Kittisak K. and Deema N. 1983. Mechanical transmission, purification and some properties of white fly borne mung bean yellow mosaic in Thailand. *Plant Disease*, 67 : 801-804.

96. Honda, Y., Iwaki M., Thongameearkom P., Kira Tiya-Angul K., Kira Tiya-Angul S., Srithongchai W., Prommin M., Kittipakorn K., Sarindu N., Deema N., Syamananda R., Hibi T. and Saito Y. 1986. Mung bean yellow mosaic isolated from mung bean in Thailand. *Technical Bulletin of the Tropical Agriculture Research Centre* No. 21 : 189-202 pp.

97. Ilias, A.M., Ali Y. and Saleem M. 1987. Incorporation of tolerance to mung bean yellow mosaic virus from local germplasm in to exotic large seeded mung bean. *II International Symposium on Mungbean, Bangkok, Thailand*, 26 pp.

98. Iwaki, M. 1978. Seed transmission of a cucumber mosaic virus in mung bean (*Vigna radiata*). *Annals of the Phytopathological Society of Japan* 44 : 337-339.

99. Iwaki, M. and Auzay H. 1977. Virus diseases of mung bean in Indonesia. *I International Mungbean Symposium, AVRDC, Taiwan* 169-172 pp.

100. Jadhav, M. R. and Sharma B.L. 1975. Field reaction of urd (*Phaseolus mungo* L.) varieties to yellow mosaic virus. *JNKVV Research Journal*, 9 : 169.

101. Jalali, B.L. Khirbat S.K. and Sangwan M.S. 1981. *In Proceedings of the National Seminar on Disease Resistance in Crop Plants, TNAU, Coimbatore,* 131 pp.

102. Kadian, O.P. 1981. Effect of some chemicals and heat on seed transmission of urd bean leaf crinkle virus. *III International Symposium of Plant Pathology, IPS, New Delhi,* 166-67 pp.

103. Kadian, O.P. 1982. Yield loss in mung bean and urd bean due to leaf crinkle disese. *Indian Phytopathology*, 35 : 642-644.

104. Kadian, O.P. 1983. Occurrence and incidence of leaf crinkle disease on urd bean and mung bean in Haryana. *Haryana Agricultural Univeristy Journal of Research* 13 : 121-126.

105. Kadian, O.P. 1983. Studies on weed plants as host range of urdbean leaf crinkle virus. *Haryana Agricultural University Journal of Research*, 13 : 602-603.

106. Kaiser, W.J. and Mossahebi G.H. 1974. Natural infections of mung bean by bean common mosaic virus. *Phytopathology,* 64 : 1209-14.

107. Kaiser, W.J., Okhovat M., Mossahebi G.H., Danesh D., Zonuzi A. and Mojahid P. 1969. Plant Pathology. *In Progress Report No. 7, Regional Pulse Improvement Project, USDA, ARS, U.S. Agency for International Development, Iran,* 47-80 pp.

108. Kassanis, B. and MaCfarlane I. 1964. Transmission of tobacco necrosis virus by zoospores of *Olpidium brassicae*. *Journal of General Microbiology*, 36 : 79-93.

109. Kaushal, R.P. and Singh B.M. 1988. Inheritence of disease resistance in black gram (*Vigna mungo*) to mung bean yellow mosaic virus. *Indian Journal of Agricultural Sciences,* 25 : 123-124.

10. Keshwal, R.L. and Khare M.N. 1987. Status of viral and mycoplasma diseases of pulses in Madhya Pradesh. *Symposium on Diseases of Oilseeds and Pulses and their Control, Agriculture College, Indore, India*, 29 pp.

11. Keshwal, R.L. and Saxena D.R. 1984. Effect of date of sowing and chemical spray on the incidence of yellow mosaic of mung bean. *Indian Phytopathology*, 37 : 411.

12. Khatri, H.L., Bhatia D.S. and Chohan J.S. 1971. Brief account of the work done on diseases of kharif pulse crops at the Department of Botany and Plant Pathology, PAU, Ludhiana during 1970-71. *Paper presented in All India Pulse Improvement Programme Workshop, Hissar, India.*

13. Khatri, R.K., Reddy P.N., Mishra A.B. and Harne S.M. 1984. Changes in phenolics, amino acids and sugar contents in mung bean due to infection by yellow mosaic virus. *Indian Phytopathology*, 37 : 408.

14. Kolte, S.J. 1971. Studies on the leaf crinkle disease of urdbean (*Phaseolus mungo* L.). *Ph.D. Thesis, U.P. Agricultural University, Pantnagar, India,* 1-64 pp.

15. Kolte, S.J. and Nene Y.L. 1970. Know the leaf crinkle disease of urd. *Indian Farmers Digest*, 3(11): 6-7.

16. Kolte, S.J. and Nene Y.L. 1972. Studies on symptoms and mode of transmission of the leaf crinkle virus of urd bean (*Phaseolus mungo* . *Indian Phytopathology*, 25 : 401-404.

17. Kolte, S.J. and Nene Y.L. 1975. Host range and properties of urd bean leaf crinkle virus. *Indian Phytopathology*, 28 : 430-431.

18. Kolte, S.J. and Nene Y.L. 1979. Urd bean (*Vigna mungo*) leaf crinkle virus.: Note worthy symptoms on host and influence of growth stages on susceptibility. *Tropical Grain Legume Bulletin* No. 15 : 5-8.

19. Kooner, B.S. and Singh Harcharan. 1980. Control of white fly and the yellow mosaic virus in green gram with granular insecticides. *Journal of Research Punjab Agricultural University,* 17 : 268-271.

20. Kooner, B.S., Singh K., Singh H. and Singh K.B. 1977. Field screening of mung bean germplasm against white fly (*Bemisia tabaci)* and yellow mosaic. *Journal of Research Punjab Agricultural University*, 14 : 75-76.

21. Kooner, B.S., Singh Harcharan and Singh K.B. 1979. Preliminary field serching of varieties of mung bean for comparative resistance to whitsfly and yellow mosaic virus *Jonnal of Research Punjab Agricultural University,* 16: 169-172.

22. Kripa Shankar and Summanwar A.S. 1975. Cytopathological studies of yellow mosaic virus disease of mung. *Indian Phytopathology* 28 : 145.

123. Lana, A.F., Egunjobi O.A. and Esuruose O.F. 1978. Studies of the soil transmission of the Nigerian okra mosaic virus. *Acta Phytopathologica Academiae Scientiarum Hungaricae,* 13 (3/4) : 307-311.

124. Legaspi, B.M., Catipon E.M. and Hubbel J.N. 1977. AVRDC Philippines outreach program mung bean studies. *I International Mungbean Symposium, AVRDC, Taiwan* 220-223 pp.

125. Lin, M.T., Anjos J.R.N. and Rios G.P. 1982. Cowpea severe mosaic virus in five legumes in central Brazil. *Plant Disease,* 66 : 57-70.

126. Liu, H.Y., Holmes F.O. and Reichmann M.E. 1969. Satellite tobacco necrosis virus from mung bean. *Phytopathology*, 59 : 833-836.

127. Majorana, G. 1966. *Phaseolus aureus* a new local lesion host of two citrus viruses. *Annali Faculty of Agriculture University Bari,* 20 : 9 pp.

128. Mali, V.R. and Vyanjane N.T. 1980. Distortion mosaic - a new virus disease of chickpea in India. *Current Science,* 49 : 599-600.

129. Mathur, R.S., Banerjee A.K. and Bajpai G.K. 1965. The effect of vector control on yellow mosaic incidence of mung in India. *Plant Disease Reporter*, 49 : 161-167.

130. McCarthy, D., Jarvis B.C. and Thomas B.I. 1970. Changes in the ribosomes extracted from mung bean infected with a strain of tobacco mosaic virus. *Journal of General Virology*, 9: 9-17.

131. McCarthy, D., Lander D.E., Hawkes S.P. and Ketteridge S.W. 1972. Effect of cycloheximide and chloramphenicol on the multiplication of tobacco necrosis virus. *Virology*, 17 : 91-97.

132. Misra, R.C,. Tripathi, D. and Sahu R.C. 1978. Yellow mosaic virus resistant mung bean for summer cultivation. *Indian Journal of Genetics and Plant Breeding*, 38 : 103-105.

133. Moses, G.J. and Nariani T.K. 1975. A mosaic disease of *Phaseolus atropurpureus. Indian Phytopathology*, 28 : 102-103.

134. Muniyappa, V., Rajeswari, R., Bharathan N., Reddy D.V.R. and Nolt B.L. 1987. Isolation and characterization of a gemini virus causing yellow mosaic disease of horse gram (*Macrotyloma uniforum* (Lam.) Verdc.) in India. *Journal of Phytopathology* 119 : 81-87.

135. Muniyappa, V., Reddy H.R., Shivashankar G. and Sreekantaradhya R. 1976. Varietal response of green gram (*Phaseolus aureus* Roxb.) to the green gram yellow mosaic disease. *Current Research* 5 : 100-101.

136. Muniyappa, V., Setty B.A.S and Shivashankar G. 1977. Reaction of *Phaseolus* species to horse gram yellow mosaic. *Indian Journal of Genetics and Plant Breeding,* 37 : 488-490.

137. Murugesan, S. and Chelliah S. 1977. Influence of sowing time on the incidence of the vector *Bemisia tabaci* and the yellow mosaic disease of green gram. *Madras Agricultural Journal*, 64 : 128-130.

138. Murugesan, S. and Chelliah S. 1977. Transmission of green gram yellow mosaic virus by the white fly *Bemisia tabaci* Genn. *Madras Agricultural Journal*, 64 : 437-441.

139. Murugesan, S. and Chelliah S. 1981. Efficacy of insecticides in the control of *Bemisia tabaci* (Genn.) a vector of the yellow mosaic virus disease on green gram. *Indian Journal of Agricultural Sciences*, 51 : 583-584.

140. Murugesan, S., Chelliah, S. and Murugesan M. 1977. Prediction of white fly vector *Bemisia tabaci* and yellow mosaic disease incidence in green gram. *Madras Agricultural Journal*, 64 : 22-28.

141. Nagraj, N.C, Muniyappa V., Satyan B.A. and Shivashankar G. 1981. Resistance to mung bean yellow mosaic virus. *Indian Journal of Genetics and Plant Breeding*, 41 : 423-425.

142. Nagraj, N.C. Muniyappa V. and Satyam B.A. 1981. Proceedings of the National Seminar in crop plants 1980, *Tamil Nadu Agriculture University Coimbatore,* 69-72 pp.

143. Nair, N.G. and Nene Y.L. 1973. Studies on the yellow mosaic of urd bean (*Phaseolus mungo* L.) caused by mung bean yellow mosaic virus. I. Transmission studeis. *Indian Journal of Farm Science* 1 : 109-110.

144. Nair, N.G. and Nene Y.L. 1973. Studies on the yellow mosaic of urid bean (*Phaseolus mungo L.*) caused by mung bean yellow mosaic virus II. Virus vector relationship. *Indian Journal of Farm Science*, 1 : 62-70.

145. Nair, N.G. and Nene Y.L. 1973. Studies on the yellow mosaic of urid bean (*Phaseolus mungo L.*) caused by mung bean yellow mosaic virus III. Factor influencing transmission and symptom expression. *Indian Journal of Farm Science,* 2 : 42-47.

146. Nair, N.G. and Nene Y.L. 1973. Studies on the yellow mosaic of urid bean (*Phaseolus mungo L.*) caused by mung bean yellow mosaic virus IV. Nature and extent of losses due to infection at various stages of growth. *Indian Journal of Farm Science*, 2 : 48-50.

147. Nair, N.G., Nene Y.L. and Naresh J. 1974 Reaction of certain urd bean varieties to yellow mosaic virus of mung bean. *Indian Phytopathology* 27 : 256-257.

148. Nalampang, A. 1977. Mung bean production in Thailand, *International Mungbean Symposium, AVRDC, Taiwan* 12-14 pp.

149. Narayanaswamy, P. and Jaganathan T. 1973. Vector transmission of black gram leaf crinkle virus. *Madras Agricultural Journal*, 60 : 651-652.

150. Narayanaswamy, P. and Jaganathan T. 1973. Sources of resistance to black gram virus diseases. *Madras Agricultural Journal* 60 : 1836-1838.

151. Narayanaswamy, P. and Jaganathan T. 1974. Effect of the virus infection on the yield components of black gram. *Madras Agricultural Journal*, 61 : 451-456.

152. Narayanaswamy, P. and Jaganathan T. 1974. Characterization of black gram leaf crinkle virus. *Madras Agricultural Journal*, 61 : 979-982.

153. Narayanaswamy, P. and Jaganathan T. 1975. Effect of black gram leaf crinkle virus infection on seed set and distribution of virus in seed. *Madras Agricultural Journal*, 62 : 151-54.

154. Narayanaswamy, P. and Jaganathan T. 1975. Studies on seed transmission of black gram leaf crinkle virus-effect of age of plant at infection. *Madras Agricultural Journal*, 62 : 287-290.

155. Narayanaswamy, P. and Jaganathan T. 1975. Seed transmission of black gram leaf crinkle virus. *Phytopathologische Zeitschrift*, 52 : 107-110.

156. Nariani, T.K. 1960. Yellow mosaic of mung (*Phaseolus aureus*). *Indian Phytopathology* 13 : 24-29.

157. Nariani, T.K. and Kandaswamy T.K. 1961. Studies on a mosaic disease of cowpea (*Vigna sinensis* Savi). *Indian Phytopathology*, 14 : 77-82.

158. Nelson, R. 1932. Investigation in the mosaic disease of bean (*Phaeolus vulgaris* L). *Michigan Agriculture Experiment Station Technical Bulletin*, 118 : 71 pp.

159. Nene, Y.L. 1969. A Survey of the viral diseases of pulse crops in Uttar Pradesh. *First Annual Report FG-In-358, U.P. Agriculture University Pantnagar*, 1-25 pp.

160. Nene, Y.L. 1969. A Survey of the viral diseases of pulse crops in Uttar Pradesh. *Second Annual Report FG-In-358, U.P. Agriculture University Pantnagar,* 1-26 pp.

161. Nene, Y.L. 1972. A Survey of the viral diseases of pulse crops in Uttar Pradesh. *Research Bulletin No. 4, G.B. Pant University of Agriculture and Technology, Pantnagar* 191 pp.

162. Nene, Y.L. 1973. Newer method of controlling yellow mosaic of moong, urd and soybean. *Indian Farmers Digest*, 6(7) : 27-28, 38.

163. Nene, Y.L. 1973. Note on a fungus parasite of *Bemisia tabaci* Genn. a vector of several plant viruses. *Indian Journal of Agricultural Sciences,* 43 : 514-516.

164. Nene, Y.L 1973. Virus diseases of some warm weather pulse crops in India. *Plant Disease Reporter*, 57 : 463-467.

165. Nene, Y.L. 1973. Control of *Bemisia tabaci* Genn., a vector of several plant viruses. *Indian Journal of Agricultural Sciences*, 43 : 433-436.

166. Nene, Y.L., Naresh J. and Nair N.G. 1971. Additional hosts of mung bean yellow mosaic virus. *Indian Phytopathology,* 24 : 415-417.

167. Nene, Y.L., Shrivastava S.K. and Naresh J.S. 1972. Evaluation of black gram (*Phaseolus mungo*) and green gram (*Phaseolus aureus)* varieties and germplasm for resistance to yellow mosaic virus of green gram. *Indian Journal of Agricultural Sciences,* 42 : 251-254.

168. Newton, W. and Peiris J.W.L. 1953. Virus diseases of plants in Ceylon. *FAO Plant Protection Bulletin*, 2(2) : 17-21

169. Nour, M.A. 1962 Witches broom and phyllody in some plants in Khartoun province, Sudan. *FAO Plant Protection Bulletin* 10 (3) : 49-56.

170. Novacky, A. and Hampton R.E. 1968. Peroxidase isoenzymes in virus infected plants. *Phytopathology*, 58 : 301-305.

171. Oblisamy, G., Devikasundaram M., Balaraman K., Purushothaman D. and Rangaswamy G. 1973. Correlation between the presence of sugars and amino acids in the leaf exudates and the quantities of microorganisms present in the phyllosphere of healthy and mosaic affected greengram. *Madras Agricultural Journal,* 60 : 481-483.

172. Pandya, B.P., Singh D.P. and Sharma B.L. 1977. Screening of mung bean (*Vigna radiata* (L.) Wilczek) germplasm for field resistance to yellow mosaic virus. *Tropical Grain Legume Bulletin*, NO. 7 : 13-14.

173. PANS. 1981. Pest Control in Tropical Grain Legumes. *PANS Manual* No. 5 : 47-125 pp.

174. Patil, N.K., Adiver S.S., Hiremath P.C. and Hegde R.K. 1987. Assessment of black gram varieties for resistance to powdery mildew and yellow mosaic virus. *Current Research*, 16(3) : 41-42.

175. Pearson, M.N. 1981. A mosaic disease of mung bean in Papua New Guinea. *Australian Plant Pathology*, 10 (2) : 31.

176. Phatak, H.C. 1974. Seed borne plant viruses-identification and diagnosis in seed health testing. *Seed Science and Technology*, 2 : 3-155.

177. Pierce, W.H. 1934. Viruses of the bean. *Phytopathology*, 24: 87-115.

178. Poehlman, J.M. 1977. What we have learned from the international mung bean nurseries. First International Mungbean Symposium, AVRDC, Taiwan, *97-100 pp.*

179. Provvidenti, R. 1986. Seed transmission of black eye cowpea mosaic virus in *Vigna mungo. Plant Disease*, 70 : 981.

180. Purivirojkul, W. and Poehlman J.M. 1977. Injury in mung bean from natural infection from cucumber mosaic virus. *Crop Science*, 17 : 654-666.

181. Purivirojkul, W., Sittiyos, P., Hsu C.H., Poehlman J.M. and Sehgal O.P. 1978. Natural infection of mung bean (*Vigna radiata*) with cucumber mosaic virus. *Plant Disease Reporter,* 62 : 530-534.

182. Rajgopalan, K., Rao M.A., Seshadri B.T. and Krishna B.A. 1981. Studies on seed transmission on Dolichos enation mosaic virus in certain legumes. *Madras Agricultural Journal*, 68 : 129-130.

183. Ram, R.S., Chenulu V.V. and Sastry L.V.S. 1984. Metabolic changes in black gram infected with mung bean yellow mosaic virus. *Indian Phytopathology*, 37 : 88-92.

184. Ramachandran, P. and Chenulu V.V. 1985. Seed health testing for seed borne viruses an urgent need. *Indian Journal of Plant Pathology*, 3 : 33-37.

185. Ramakrishnan, G., Kandaswamy T.K., Damodaran A.P.S. and Ayyavoo R. 1973. Studies on new mosaic viruses occurring on *Phaseolus aureus* Roxb. *Madras Agricultural Journal*, 60 : 465-468.

186. Ramakrishnan, G., Ranganathan K., Murugesan S., Damodaran A.P.S., Janaki I. and Ayyavoo R. 1973. Studies on mosaic disease of black gram (*Phaseolus mungo* L.). *Madras Agricultural Journal*, 60 : 469-471.

187. Ramakrishnan, G., Rangaraju R., Thangamani G., Sarojini D.A. and Kandaswamy T.K. 1969. Studies on a new mosaic disease of *Phaseolus aureus* L. *Madras Agricultural Journal*, 56 : 378-80.

188. Ramanujam, S. 1977. Grain legumes in India. *Vienna, Austria International Atomic Energy Agency*, 187 pp.

189. Ramarao, M.A. and Rajgopalan K. 1987. Influence of Dolichos enation mosaic virus infection on pod characters and yield in different varieties of black gram and green gram. *Journal of Research Andhra Pradesh Agricultural University*, 15 : 66-68.

190. Rao, M.A. 1973. Studies on the additional hosts (black *gram* and green gram) for Dolichos enation mosaic virus (DEMV). Thesis, S.V. Agriculture College Tirupati, India, 72 pp.

191. Rao, M.A. and Rajgopalan K. 1984. Rare phenomenon of abnormal elongation of infloresence in green gram (*Phaseolus aureus*) varieties infected by Dolichos enation mosaic virus. *Current Science* 53 : 546-547.

192. Rathi, Y.P.S. and Nene Y.L. 1974. Two additional hosts of mung bean yellow mosaic virus. *Indian Phytopathology*, 27 : 429-430.

193. Rathi, Y.P.S. and Nene Y.L. 1974. Some aspects of the relationship between mung bean yellow mosaic virus and its vector *Bemisia tabaci*. *Indian Phytopathology*, 27 : 459-462.

194. Rathi, Y.P.S. and Tripathi H.S. 1989. Management of mung bean yellow mosaic and pigeonpea sterility mosaic. *Indian Phytopathology*, 42 : 314-315.

195. Rathi, Y.P.S. and Nene Y.L. 1975. Sex of *Bemisia tabaci* Genn. in relation to the transmission of mung bean yellow mosaic virus. *Acta Botanica Indica*, 2 (1) 74-76.

196. Rathi, Y.P.S. and Nene Y.L. 1976. Influence of different host combinations on virus vector relation of mung-bean yellow mosaic virus. *Pantnagar Journal of Research*, 1 : 107-111.

197. Raychaudhury, S.P., Dhingra K.L. and Varma J.P. 1977. Virus diseases of pulse crops in India. *Seed and Farms*, 3(4) : 7-17.

198. Raychaudhury, S.P.and Pathanian P.S. 1950. A mosaic disease of *Crotalaria mucronata* Desv. (*C. striata* DC.). *Current Science,* 19 : 213.

199. Reddy, P.C. and Krishnaiah K. 1983. Assessment of yield losses due to leaf crinkle virus on black gram. *Indian Journal of Mycology and Plant Pathology,* 13 : 19.

200. Regupathy,A., Rathnaswamy R., Venkatarayan D. and Subramanian T.R. 1975. Physiology of yellow mosaic virus in green gram (*Phaseolus aureus*) with reference to its preference by *Empoasca kerri* Pruthi. *Current Science*, 44 : 577-578.

201. Saleh, N., Honda Y., Iwaki M. and Tantera D.M. 1986. Occurrence of black gram mottle virus on mung bean in Indonesia and seed transmission of the virus. *Technical Bulletin of the Tropical Agricultural Research Centre,* No. 21 : 203-212.

202. Samuel, C.S., Janarthanam R., Kandaswamy T.K. and Subramanian K.S. 1972. New hosts for the yellow mosaic virus of *Dolichos lablab. Current Science*, 41 : 686-687.

203. Sandhu, T.S. 1977. Breeding for yellow mosaic virus resistance in mung bean. *I International Mungbean Symposium AVRDC, Taiwan* 176-179 pp.

204. Sandhu, T.S., Brar J.S., Sandhu S.S. and Verma M.M. 1985. Inheritence of resistance to mungbean yellow mosaic virus in green gram. *Journal of Research Punjab Agricultural University*, 22 : 607-611.

205. Scott, H.A. and Phatak H.C. 1979. Properties of black gram mottle virus. *Phytopathology* 69 : 346-348.

206. Sen Gupta, P.K. . 1974. Diseases of major pulse crops in India. *PANS,* 20 : 409-415.

207. Seth, M.L., Raychaudhury S.P. and Nath R. 1967. A new mosaic disease of brinjal (*Solanum melongena* L.). *Phytopathologische Zeitschrift*, 59 : 385-389.

208. Shahare, K.C. and Raychaudhury S.P. 1963. Mosaic disease of urd (*Phaseolus mungo* L.). *Indian Phytopathology,*16 : 316-18.

209. Shakoor, A., Ahsanul Haq M., Sadiq M. and Sarwar G. 1977. Proceedings of a symposium on the use of induced mutations for improving disease resistance in crop plants. *Vienna, Austria, International Atomic Energy Agency,* 293-302 pp.

210. Shanta, P. and Menon K.P.V. 1961. Studies on some properties of coconut wilt virus. *Indian Coconut Journal*, 15 : 36-46.

211. Sharma, Indu and Dubey G.S.1981. Studies on seed borne nature of aphid transmitted urd bean leaf crinkle virus. *Third International Symposium of Plant Pathology, I.P.S., New Delhi*, 166 pp.

212. Sharma, Indu and Dubey G.S. 1981 Histopathology of leaf crinkle virus infected urd bean plants. *Indian Phytopathology*, 34 : 113.

213. Sharma Indu and Dubey G.S. 1983. Histological changes in urd bean flowers due to urd bean leaf crinkle virus infection. *Zeitschrift fur Pflanzenkrankheiten und Pflanzenschutz*, 90 : 63-67.

214. Sharma, Indu and Dubey G.S. 1984. Control of urd bean leaf crinkle virus through heat treatment, chemotherapy and resistance. *Indian Phytopathology*, 37 : 26-30.

215. Sharma, Indu and Dubey G.S. 1985. Some pathphysiological changes in urd bean (*Vigna mungo*) induced by urd bean leaf crinkle virus. *Indian Phytopathology*, 38 : 624.

216. Sharma, Indu and Dubey G.S. 1985. Anatomy of urd bean leaf crinkle virus infected urd bean root. *Indian Journal of Plant Pathology*, 3 : 236-237.

217. Shivanathan, P. 1977. A seed borne virus of *Phaseolus aureus* Roxb. *Symposium on Virus Diseases of Tropical Crops, Tropical Agriculture Research Series* No. 10 : 143-150.

218. Shukla, G.P. 1986 Etiology and breeding for resistance to yellow mosaic in greengram-A reivew. *Agriculture Review*, 7 (1) : 31-38.

219. Shukla, G.P. and Pandya B.P. 1985. Resistance to yellow mosaic in green gram. *SABARAO Journal*, 17 : 165-171.

220. Shukla, G.P., Pandya B.P. and Singh D.P. 1978. Inheritence of resistance to yellow mosaic in mung bean. *Indian Journal of Genetics and Plant Breeding*. 38 : 357-360.

221. Shyam, R. and Bhatnagar P.S. 1965. Phyllody in black gram (*Phaseolus mungo* L.). *Science and Culture,* 31 : 312 - 313.

222. Singh, A.K. and Singh R.B. 1980. Alteration in protein yield and protein quality in urd bean infected by southern bean mosaic virus. *Annales de Phytopathologie*, 12 : 131-138.

223. Singh, A.K. and Srivastava S.K. 1985. Effect of urd bean mosaic virus infection on the yield and chemical composition of urd bean fruits. *Indian Phytopathology*, 38 : 85-89.

224. Singh, A. K. and Srivastava S.K. 1985. Nodular physiology of urd bean as affected by urd bean mosaic virus. V. Effect of some enzymatic —activity. *Phyton Austria,* 25 : 213 - 217.

225. Singh, A.K., Srivastava S.K. and Singh A.K. 1985. Effect of milk on infectivity of urd bean mosaic. *Indian Journal of Mycology and Plant Pathology,* 15 : 311-312.

226. Singh, A.K., Srivastava S.K. and Singh S.V. 1985. Nodular physiology of urd bean affected by urd bean mosaic virus. VI. Leghaemoglobin content. *Acta Botanica Indica*, 13 : 131-133.

227. Singh, A.K., Srivastava S.K. and Singh A.K. 1985. Nodular physiology of urd bean as affected by urd bean mosaic virus. II. Effect on nitrogen content. *Indian Journal of Virology*, 1 : 325-329.

228. Singh, B.R., Singh M., Yadav M.D. and Dingar S.M. 1982. Yield loss in mung bean due to yellow mosaic. *Science and Culture,* 48 : 435-436.

229. Singh, B.R., Singh R.B., Singh Harcharan and Gangulu R. 1979. Observations on a virus disease of *Coffee seuna*. *Indian Journal of Mycology and Plant Pathology*, 9 : 101.

230. Singh, B.V. and Ahuja M.R. 1977. *Phaseolus sublobatus* Roxb., a source of resistance to yellow mosaic virus for cultivated mung. *Indian Journal of Genetics and Plant Breeding,* 37 : 130-132.

231. Singh, D. and Patel P.N. 1977. Studies on resistance of crops to bacterial diseases in India. VIII. Investigations on inheritance of resistance to bacterial leaf spot and yellow mosaic diseases and linkage, if any, with other characters of mung bean. *Indian Phytopathology*, 30 : 202- 206.

232. Singh, Dharam Pal. 1954. Phyllody of mung. *Science and Culture,* 20: 194.

233. Singh, D.P. 1980. Inheritance of resistance to yellow mosaic virus in black gram (*Vigna mungo* (L.) Hepper). *Theoritical and Applied Genetics,* 57 : 233-235.

234. Singh, D.P. and Sharma B.L. 1981. Evaluation of mung bean germplasm. *Madras Agricultural Journal*, 68 : 289-295.

235. Singh, Gurdip, Singh K., Gill A.S. and Chhabra K.S. 1981. Proceedings of the National Seminar on Disease Resistance in crop plants, 1980. Tamil nadu Agriculture University Coimbatore, 72-76 pp.

236. Singh, Gurdip, Singh K., Gill A.S., Kapoor Shashi and Verma M.M. 1983. Screening for resistance to mung bean yellow mosaic virus in urd bean (*Vigna mungo* (L.) Hepper). *Indian Phytopathology*, 36 : 185.

237. Singh, Gurdip, Kapoor S. and Singh Kuldip. 1987. Multiple disease resistance in mung bean with special emphasis on yellow mosaic virus. *Second International Symposium on Mungbean, Bangkok, Thailand*, 25-26 pp.

238. Singh, J.P. 1980. Effect of virus diseases on growth components and yield of mung bean (*Vigna radiata*) and urd bean (*Vigna mungo*). *Indian Phytopathology*, 33 : 405- 408.

239. Singh, J.P., Kadian O.P. and Verma J.P. 1979. Survey of virus diseases of mung baen and urd bean in Haryana. *Haryana Agrigultural University Journal of Research*, 9 : 345-347.

240. Singh, J.P. and Suhag L.S. 1982. Pigment, nucleic acid and protein concentration in the virus infected mung bean and urd bean leaves. *Indian Journal of Mycology and Plant Pathology*, 12 : 61-63.

241. Singh, J.P. and Varma J.P. 1977. Mosaic disease of mung and urd bean in Haryana. *Proceedings of National Academy of Sciences India B*, 47 : 33-40.

242. Singh, J.P. and Varma J.P. 1977. Effect of some insecticides and oils on incidence of yellow mosaic and grain yield of mung bean cv. Varsha. *Proceedings of the National Academy of Sciences India B*, 47 : 219-225.

243. Singh, R. and Mall T.P. 1973. Changes in agronomic characters and contents of mung bean (*Phaseolus aureus* Roxb.) infected with arhar mosaic virus. *Acta Agronomica Universidad National de colombia*, 23 : 51-65.

244. Singh, R. and Mall T.P. 1973. Studies on the nodulation and nitrogen fixation by infected leguminous plants. I. Effect of arhar mosaic virus infection on nitrogen value, nodulation and nitrogen fixation by some pulse crops. *Plant and Soil*, 41 : 279 - 286.

245. Singh, R. and Mall T.P. 1974. Physiology of *Phaseolus mungo* L. affected with urid mosaic virus. I. Effect on chlorophyll content, catalase and peroxidase activity. *Portugaliae Acta Biologica*, 13 : 63-71.

246. Singh, R. and Mall T.P. 1974. Physiology of *Phaseolus mungo* L. affected by urd bean mosaic virus. II. Effect on carbohydrate and phosphorus. *Phyton Austria*, 15 : 235-238.

247. Singh, R. and Chowdhury J.R. 1983. Influence of the infection by common bean mosaic virus on nodulation and nitrogen fixation by mung bean (*Vigna radiata* (L.) Wilczek). *Rivista de Agricoltura Subtropicale,* 77 : 249 - 257.

248. Singh, R. and Singh R. 1975. Studies on a mosaic disease of urid bean (*Phaseolus mungo*). *Phytopathologica Mediterranea,* 14 : 55-59.

249. Singh, R. and Singh R. 1978. Studies on mosaic causing disease of urd bean. *Indian Journal of Mycology and Plant Pathology*, 8 : 185-187.

250. Singh R. and Singh R.B. 1978. Changes in enzymatic activity of mung bean leaves due to mung bean severe mosaic virus infection. *Legume Research,* 2 : 53-55.

251. Singh, R. and Singh R.B. 1979. Changes in chlorophyll content, primary productivity and carbohydrate fractions of mung bean due to mung bean severe mosaic virus infection. *Iranian Journal of Plant Pathology*, 15(1/4) : 55-63, 83.

252. Singh, R. and Singh R.B. 1983. Changes in chemical composition of mung bean fruits due to mung bean severe mosaic virus infection. *Indian Phytopathology*, 36 : 129-131.

253. Singh, R. and Srivastava R.S. 1983. Effect of nitrogen nutrition and Rhizobium on multiplication and symptom expression of common bean mosaic virus in mung bean. *Zeitschrift fur Pflanzenkrankheiten und Pflanzenschutz,* 90 : 207-212.

254. Singh. R.A., Amin K.S. and Gurha S.N. 1986. Sources of resistance of yellow mosaic virus in mung bean and urd bean, *Proceedings National Seminar on Whitefly Transmitted Plant Virus Diseases, IARI, New Delhi*, 35 pp.

255. Singh, R.A. and Gurha S.N. 1980. Problems of some important pulses and prospects of their control. *Pesticides Annual.* 81-85 pp.

256. Singh. R.A., Gurha S.N., Misra D.P. and Gangal L.K. 1980. Role of systemic insecticides in augmenting yield with reduced yellow mosaic virus incidence in mung bean. *Indian Journal of Plant Protection,* 8 : 167 - 169.

257. Singh, R.A., Sahambi H.S. and Gurha S.N. 1982. Investigations on yellow mosaic of urd bean (*Vigna mungo*) with particular reference to yield losses and sources of resistance. *Indian Journal of Mycology and Plant Pathology*, 12:115.

258. Singh, R.N. 1976. Leaf curl disease of bean. *FAO Plant Protection Bulletin*, 24 : 100-101.

259. Singh, R.N. 1976. Studies on the mosaic mottle disease of urd bean (*Phaseolus mungo* L.). *Pesticides*, 10 (4) : 56.

260. Singh, R.N. 1976. Natural infection of mung bean by bean common mosaic virus in India. *Indian Journal of Mycology and Plant Pathology*, 6 : 94-95.

261. Singh, R.N., Nene Y.L. 1977. Factors affecting symptom expression and incubation period of bean common mosaic virus in urd bean. *Indian Journal of Mycology and Plant Pathology*, 7 : 111-114.

262. Singh, R.N. and Nene Y.L. 1977. Diagnostic symptoms of mosaic mottle disease of urid bean (*Phaseolus mungo* L.). *Seeds and Farms*, 3 (3) : 23-25.

263. Singh, R.N. and Nene Y.L. 1978. Further studies on the mosaic mottle disease of urd bean. *Indian Phytopathology*, 31 : 159-162.

264. Singh, R.N., Nene Y.L. and Srivastava S.K. 1983. Evaluation of urd bean and mung bean genotypes for resistance to bean common mosaic virus and losses caused by the disease. *Legume Research*, 6 : 69-73.

265. Singh, Shamser, Bisht I.S. and Dabas B.S. 1987. Screening of urd bean (*Vigna mungo*) germplasm for field resistance to yellow mosaic virus. *Indian Journal of Mycology and Plant Pathology*, 17 : 216-217.

266. Sittiyas, P., Poehlman J.M. and Sehgal O.P. 1979. Inheritance of resistance to CMV infection in mung bean. *Crop Science*, 19 : 51-53.

267. Sivaprakasam, K., Anabalagan R., Rajendran R. and Radhakrishnan T. 1983. Influence of N, P.K, nutrition on incidence of yellow mosaic in urd bean. *Pulse Crops Newsletter*, 3 : 55-56.

268. Sivaprakasam, K., Jagadeesan M., Kannapiran C. 1976. Mineral content in the healthy and yellow mosaic infected black gram. *Annamalai University Agricultural Research Annual*, 6 : 171 : 173.

269. Sivaprakasam, K., Jagadeesan M., Kannapiran C. and Pillayarsamy K. 1976. Changes in the mineral content of black gram leaves due to leaf crinkle virus. *Madras Agricultural Journal*, 63 : 123-125.

270. Sivaprakasam, K., Pillayarsamy K., Rajamani A. and Souminirajagopalan C.K. 1974. Evaluation of black gram (*Phaseolus mungo* L.) and green gram (*Phaseolus aureus* Roxb.) varieties for resistance to yellow mosaic virus of green gram. *Madras Agricultural Journal,* 61 : 1021-1022.

271. Srivastava, K.M., Varma G.S. and Varma H.N. 1969. A mosaic disease of black gram (*Phaseolus mungo*). *Science and Culture,* 35 : 475-476.

272. Teakle, D.S. 1962. Transmission of tobacco necrosis virus by a fungus *Olpidium brassicae. Virology,* 18 : 224-231.

273. Teakle, D.S. 1973. Use of local lesion method to study the effect of celite and inhibitors on virus infection of roots. *Phytopathologische Zeitschrift,* 77 : 209-215.

274. Teakle, D.S. and Hiruli C. 1966. Vector specificity in *Olpidium. Virology,* 24 : 539-544.

275. Teakle, D.S. and Yarwood C.E. 1962. Improved recovery of tobacco necrosis virus from roots by means of *Olpidium brassicae. Phytopathology*, 52 : 359-369.

276. Thomas, H. R. and Zaumeyer W.J. 1950. Red node, a virus disease of beans. *Phytopathology,* 40 : 832-846.

277. Thongameearkom, P., Honda Y., Saito Y. and Syamananda R. 1981. Nuclear ultrastrutural changes and aggregates of virus like particles in mung bean cells affected by mungbean yellow mosaic disease. *Phytopathology*, 71 : 41-44.

278. Thongameearkom, P., Kittipakorn K. and Surin P. 1981. Out break of mung bean yellow mosaic disease in Thailand. *Thai Journal of Agricultural Sciences,* 14 : 201 - 206.

279. Tolmocheva, E.A. 1966. Viruses of legumes in the central botanical garden of Byclorussian academy of sciences. *Vestsi Akad. Navuk BSSR, Ser Biyal. Navuk,* 4 : 125 - 129.

280. Tripathi, R.K.R. and Tripathi G. 1983. Assay on NR activity in *Vigna radiata* infected by broad bean mosaic virus (BBMV). *Legume Research,* 6 : 43 - 44.

281. Tripathi, R.K.R., Tripathi G., Singh K.B. and Kumar S. 1983. Effect of broad bean mosaic virus on nitrate reductase activity in *Vigna mungo* cv. Type 9. *National Academy of Science Letter,* 6 : 41-42.

282. Tripathi, R.K., Vohra K. and Beniwal S.P.S. 1975. Changes in phenoloxidase and peroxidase activities and peroxidase isoenzymes in yellow mosaic virus infected mung bean (*Phaseolus aureus* L.). *Indian Journal of Experimental Biology*, 13 : 513-514.

283. Tsuchizaki, T. 1986. Partial characterization and serological relationship of 3 poty viruses isolated from leguminous crops in Thailand. *International Symposium on Virus Diseases of Rice and Leguminous Crops in Tropics, Japan,* 115-120 pp.

284. Tsuchizaki, T., Iwaki M., Thongameearkom P., Sarindu N. and Deema N. 1986. Bean common mosaic virus isolated from mung bean (*Vigna radiata*) in Thailand. *Technical Bulletin of the Tropical Agriculture Research Centre,* No. 21 : 184-188.

285. Van Hall, C.J.J. 1921. Diseases and pests of cultivated plants in the Dutch East Indies during 1921. *Meded. Institute Voor Plantenzeikten,* 53 : 46.

286. Verma, Anupam, 1985. Natural occurrence of cowpea mild mottle virus in *Vigna mungo. Indian Phtopathology* 38 : 626.

287. Varma, A. and Subramanyam K. 1986. National Seminar on Whitefly Transmitted plant Virus Diseases, June 1986, IARI, New Delhi, 46 p.

288. Varma, J.P., Kadian O.P. and Singh J.P. 1973. Occurrence and incidence of virus diseases of mung and urd in Haryana. *Indian Phytopathology*, 26 : 592-594.

289. Vasudeva, R.S. 1958. Report of the Division of Mycology and Plant Pathology. *Scientific Report, Agriculture Research Institute New Delhi, 1955-56*, 85-104 pp.

290. Venkataraman, S. and Subbarao N.S. 1974. Variation of *Rhizobium* sp. (cowpea group) from root nodules of healthy and yellow mosaic virus infected *Phaseolus aureus* plants. *Phytopathologische Zeitschrift,* 80 : 29-34.

291. Verma, H.N. and Kumar V. 1982. Prevention of plant virus disease by *Mirabilis jalopa* leaf extract. *New Botanist,* 7 : 87-91.

292. Verma, H.N., Rastogi P., Prasad V. and Srivastava A. 1985. Possible control of natural virus infection of *Vigna radiata* and *Vigna mungo* by plant extracts. *Indian Journal of Plant Pathology,* 3 : 21-24.

293. Verma, R.P.S. and Singh D.P. 1986. The allelic relationship of genes giving resistance to mung bean yellow mosaic virus in blackgram. *Theoritical and Applied Genetics,* 72 : 737-738.

294. Vidhyasekaran, P., Arjunan G. and Ranganathan K. 1976. Field tolerance of some green gram types to yellow mosaic. *Madras Agricultural Journal,* 63 : 54-55.

295. Vidhyasekaran, P. and Kandaswamy D. 1972. Carbohydrate metabolism of *Phaseolus aureus* infected with obligate and facultative parasites. *Indian Phytopathology*, 25 : 48 - 54.

296. Virmani, S.S., Singh K.B. and Singh K. 1976. Note on the screening of green gram germplasm against yellow mosaic virus disease in India. *Indian Journal of Agricultural Sciences,* 46 : 243-245.

297. Vohra, K. and Beniwal S.P.S. 1979. Effect of mung bean yellow mosaic virus on yield and seed quality of urd bean (*Vigna mungo*). *Seed Research,* 7 : 168-174.

298. Williams, F.J., Grewal J.S. and Amin K.S. 1968. Serious and new diseases of pulse crops in India in 1967. *Plant Disease Reporter,* 52 : 300-304.

299. Yang, C.Y. 1977. Mung bean diseases and control. *First International Mungbean Symposium, AVRDC, Taiwan,* 141-146 pp.

300. Yein, B.R. and Singh Harcharan. 1982. Effect of pesticides and fertilizers on the population of whitefly and incidence of yellow mosaic virus in green gram. *Indian Journal of Agricultural Sciences*, 52 : 852-855.

301. Kadian, O.P. 1989. Effect of environment on incidence and development of leaf crinkle disease in uridbean. *Indian Phytopathology* 42 : 272.

302. Prasada Rao, R.D.V. and Chakrabarty S.K. 1989. Occurrence of whitefly transmitted virus causing rugose symptoms on uridbean in Andhra Pradesh. *Indian Phytopathology* 42 : 278.

CHAPTER 5
NEMATODE DISEASES AND MITES

Legumes are the good hosts of nematodes and they are attacked severely in warm areas. The greengram and blackgram are attacked by species of *Meloidogyne, Heterodera, Rotylenchulus, Pratylenchus* and several others but first three cause much damage to these crops.

5.1. ROOT KNOT NEMATODE

Greengram and blackgram were reported highly susceptible to *Meloidogyne incognita* from tobacco causing root knot symptoms (29) in India. The disease was observed occurring on mungbean in Philippines where pathogenicity of the nematode was confirmed (7, 20). In India the root knot on greengram and blackgram under natural conditions was reported in 1972 by Singh (50) in Uttar Pradesh and later it was observed in Orissa (16), Gujrat (41) and Madhya Pradesh (1). The root knot on mungbean has also been reported from Alabama(42). In India, the susceptible varieties of greengram and blackgram are reported to bear 60-100 % less pods resulting in 70-80 % reduction in yield (50). The avoidable yield losses of 29-54 % (average 39%) have been worked out in different varieties of mungbean suffering from root knot nematode (24). In Philippines, the reduction in yield of mungbean due to *Meloidogyne acrita* has been reported 28 and 40 % under field and pot conditions respectively but it was more than 60% when mixed population of *M. acrita* and *Rotylenchulus reniformis* was used (10).

Symptoms :

Different species of *Meloidogyne* cause more or less similar symptoms on mungbean and urdbean. The symptoms produced by them are seen on root as well as on foliage.

Below ground symptoms : The infective second stage juveniles invade the young roots and initiate the hyperplasia and hypertrophy of root tissues surrounding feeding sites. It results in the formation of characteristic root galls (Fig. 5.1). In case of mild infections, discrete galls are formed along the roots specially the lateral roots, that can be mistaken for root nodules. If the infection is severe then, along with larger galls, the root growth is reduced showing necrosis of various stages (11, 50). If the nematode population is very high then deterioration of root starts and the plant does not revive (27). The

Fig.5.1.
Root knot on green gram (Courtesy : Dr. D. J. Patel)

injuries thus caused, serve the entry points for secondary invaders such as fungi and bacteria which may amplify the damage (11). Inside the vascular cylinder, giant cells are formed at feeding sites which block and disturb the vascular tissues adversely affecting the uptake of water and nutrients from the soil (11).

Above ground symptoms : A massive nematode invasion of newly germinated seedlings may cause failure of seedling emergence. Plants attacked are deprived of normally functioning root system and suffer from mineral and moisture defficiency showing leaf chlorosis during the moisture stress conditions but leaf may turn green after irrigation if the infection is mild. If the infection is severe or if it occurs early then chlorosis is followed by plant stunting, drying of leaf margins before flowering and ultimately death of the plant (10, 41). The leaves may shed (25) and in some cases early flowering may be induced with only few small pods (10). The above ground symptoms appear in clearly defined patches in the field and are more obvious in adverse growing conditions such as sandy soil or the drought (38).

Causal organism :

Meloidogyne incognita (Kofoid & White) Chitwood.

Meloidogyne javanica (Troub) Chitwood

Meloidogyne acrita (= *M. incognita acrita* Chitwood)

Meloidogyne arenaria (Neal) Chitwood

Meloidogyne hapla Chitwood

The greengram is infected by all the species except *M. hapla* (11, 38, 41) while blackgram is infected by *M. incognita, M. javanica* and *M. hapla* (38).

The root galls contain pearly white swollen adult females which are 0.5 to about 0.8 mm long. The body tapers anteriorly to a narrow neck and mobile head with stylet, relatively massive median bulb and large oesophageal glands. An egg sac often protrudes posteriorly from the female to the exterior of the gall. It usually becomes brown and contains large number of eggs. At the centre of the posterior region of the female, the cuticle has a pattern of the cuticular marking surrounding the anus and the vulva. This pattern is used for the identification of the species. The eggs hatch without requiring any root diffusate producing second stage juveniles which invade the root at or close to the tip and migrate to the site of differentiating vascular tissues. Here, several giant cells develop around the head of the nematode on which they feed. The nematode enlarge to a sausage shape, moults three times and becomes the adult male or female. The male is elongate, vermiform, about 1.1 mm long and the posterior characteristically twisted through 90 degrees or more. The male leaves the root, enters the egg sac and mates with the sedentary female. Parthenogenesis is commonly observed in *Meloidogyne* spp. The time required for completing the life cycle depends on the species of the parasite, the host and the environmental conditions. The host influences the size of the nematode also. The females of *M. incognita* produced in the highly susceptible hosts (mungbean, urdbean and mothbean) are reported to be larger and robust while they are smaller and slender when produced in resistant hosts like soybean, cowpea and pigeonpea (39). The host range of root knot nematodes is very wide covering the plant species of several families.

Effect on host :

Invasion of the mung bean roots by the *Meloidogyne* spp. hampers the nitrogen fixing process by the *Rhizobium sp.* It reduces the number of nodules in primary and secondary roots and the nitrogen contents of root and shoot ultimately reducing the growth of plant (5, 25, 45). The giant cells associated with the adult females of *Meloidogyne* spp. are found metabolicaly active. The crushed phloem cells, various stages of the nematode, the egg mass and parenchyma cells lining the cavity containing the female nemetode have been reported to be rich in ascorbic acid (19) and it exhibit some relation with the resistance of the host.

Host resistance :

Large number of greengram and blackgram entries have been reported to possess resistance against root knot nematode (Table 5.1).

Table 5.1

Entries of greengram and blackgram resistant to root knot nematodes.

Nematode species	Entires tested	Entries showing resistance	Reference
		GREENGRAM	
Meloidogyne incognita	-	T44 (immune), T 2 (tolerant)	(31)
" "	277	53P 61, 68P 81, 91P 131-68, 125P 179-69, 160P 68, 243P 362-68, 244P 368-68, 478Pb 297, 479Pb 294, 218P 330, NP 18, RS 4	(26)
" "		ML 3, ML 62, PLM 141	(2)
" "	24	ML 5, ML 62, ML 80, K 851, T 44, TAP 7, TT 8E, TT 9E, PDM 14, Hyb 12-4, ML 3, ML 70	(17)
" "	30	MH 309, K 100B, K 1125, Pusa 8732	(48)
Meloidogyne javanica	-	ML 50	(2)
" "	-	DMG 1052-1, 1052-2, DMG 1050-1, DMG 1074-2 (all	

(Contd...)

		resistant with good yield); 752/LM 439, DMG 1096, DMG 1074-2 (as resistant donner);	(24)
M.incognita & *M. Javanica*	27	B 105, ML 80, Pant M-2 were moderately resistant.	(15)
	13	Gujrat 1, ML 3 and ML 62 were less susceptible	(41)
M. Incognita, *M. javanica* and *M. acrita*	several hundred	A multifoliate mutant was resistant *but not to M. acrita*	(11)
		BLACKGRAM	
M.incognita		Mash 1-1, No 41-43, Pusa 1	(27)

Some of the greengram varieties (238/V-4406, LM 515B] give good yield inspite being susceptible to root knot nematode *M. incognita* (24). The reaction of greengram varieties has been reported to differ greatly in some cases. For example, ML 3, and ML 62 were found susceptible in Gujrat (41) but they showed resistant or moderately resistant reaction at other places (2, 17). This difference may be due to existence of different pathotypes of the causal organism. The resistance against the root knot nematode has been considred as characterized by reduced larval population, fewer shrunken giant cells at the feeding sites, reduced growth of the invading nematode and retarded life cycle of the parasite (26). The average gall index, egg mass per plant or number of eggs or larvae/plant are also the important factors indicating resistance against the parasite (17, 48). Based on these characters, scale has been developed for rating the severity of the infection by root knot nematode as used by Krishna Murthy *et al.* (29) :

0 = No infection, larvae if enter roots do not develop in to egg laying females.

1 = Extremely light infection with only occasional matured females with egg masses.

2 = Light infection with matured females and egg masses easily seen.

3 = Moderate infection with full grown females and egg masses moderately abundant.

4 = Severe infection with matured females and egg masses very abundant.

The resistance has been shown to be related with the chemical contents of the host plant particularly amino acids. Resistant entries of mungbean are reported to contain more argenine, leucine / norleucine and threonine as compared to the susceptible ones (26).

Control :

Chemical : Application of temik @ 10 kg/ha (10 % a.i.) caused recovery of the blackgram plants infected with *M.incognita* in India (27). The control of root knot nematode with 12.4 % increase in yield of mungbean has been reported by soil fumigation with vyadate @ 4.58 l a.i./ha (30). Soil application of aldicarb @ 0.5 kg a.i. per hectare controls the root knot nematode increasing the yield by 3.5 times and the results are a pointer to the likely associations between the pests resulting in the disease severity and crop losses (43). The soil application of aldicarb 1.5 kg/ha alone or with 12.5 kg N/ha and 40 kg P_2O_5 per hectare reduces the population of *M. incognita* as well as the root galling by the parasite and increases the yield of mungbean (56).

Crop rotation : Monoculture of mungbean or any other crop susceptible to nematode increases the population of the parasite and the damage caused to mungbean. The population is greatly reduced by atleast 2 successive croppings of corn or *Tagetis erecta* or their use as intercrops with mungbean or three clean fallow periods (1). In well drained fields, one or two croppings of paddy or maize preceding a susceptible crop during each cropping year gives fair nematode control (8, 11). Crops like *Sesamum indicum* or *Crotalaria juncea* also help in reducing the population of root knot nematodes.

Flooding : In Philippines, more than 50 % upland fields are found to have the population of *M. acrita* at the time of harvest (9). In upland areas, the mixed population of *M. acrita* and *Rotylenchulus reniformis* is present in almost every field where mungbean is grown and heavy root infestations are frequently associated with crop decline. In lowland areas where mungbean usually follows rice, very low population of nematode is encountered (11). It indicates the utility of flooding in nematode control. The flooding controls the population of nematodes in rice-field legumes-fallow system fields. The lowest population is observed in 2-3 months/year flooding (8). Warm conditions reduce the period of flooding required for control of root knot nematodes and the larvae are more easily killed than the eggs.

Soil amendments : The infection of root knot nematodes on green gram is observed more severe in light soil conditions in Gujrat, India (41). It is more severe in sandy loam soil especially under monoculture conditions and in irrigated crops in absence of rains during the rainy season (24). The severity and the population of root knot nematodes is greatly reduced by soil application of chicken dung or the rice straw compost in case of mungbean (11).

To reduce the damage, it is necessary to control other root invaders particularly fungi and bacteria which enhance the damage many fold. The borers are also the important factor in increasing the damage caused by nematodes (43).

5.2. RENIFORM NEMATODE :

Reniform nematode is commonly found in tropics and has a very wide host range. It is found associated with root rot and decay of mungbean in Philippines (6). It causes 30 % reduction in growth of the mungbean plant (4) and in combination with root knot nematodes causes 24 % yield losses (10). Under inoculated conditions, the losses may be more than 60 per cent. The nematode infects blackgram also (22). The reduction in yield is due to the adverse effect of the nematode on different growth attributes of urd and mungbean (37). The infection of greengram @ 1000 or more nematodes per 500 g soil significantly reduces the root weight and nodulation (23). It delays the emergence of greengram seedlings by 7-9 days at 1 nematode/g soil and reduces the seedling stand considerably (36). The reniform nematode infects the pulse crops in Madhya Pradesh, India also (1).

Symptoms :

The immature migratory females are the infective stages of the reniform nematode. They invade roots and create a permanent feeding site in the phloem or the pericycle causing alteration or disruption of cells of the vascular cylinder (38). The infected roots are usually necrotic and the severity increases with the infection level in mungbean (11). Nematode feeding destroys the epidermal cells, causes discolouration and death of cortical tissues. It is not associated with any obvious symptom of root or the foliage damage but causes an overall reduction in growth and yield of the plant. Some times yellowing and chlorosis of the foliage is observed under moisture stress conditions (38).

Causal organism :

Rotylenchulus reniformis Linford and Oliveria 1940

It is a semi-endoparasitic pest of many crops including mung bean and urdbean. The second stage juveniles hatch from eggs in soil and undergo 3 moults in succession without feeding. The immature adult females penetrate roots and push through cortex cells until their heads reach in endodermal cells. The broken cortex cells constrict the developing female so that the mature female has a knobbed anterior portion within the cortex and a swollen posterior outside the root. The female deposits a large mass of mucoid egg sac material in to which eggs are laid. The motile males mate with females which become obese and lay eggs. The soil particles tend to adhere to the gelatinous egg masses so that the infested root appears darker than the healthy. The reproduction potential of the nematode is maximum at 100 nematodes per 500 g soil and thus is density dependent (23).

Host resistance :

Resistance in mungbean as well as urdbean varieties has been identified against *R. reniformis* (Table 5.2).

Table 5.2
Entries of greengram and blackgram resistant to *Rotylenchulus reniformis*.

Crop	Entries tested	Entries showing resistance	Reference
Greengram	33	ML 56, Pusa 105 were resistant; ML 5, PDM 54, PLM 572, PS 16, MUG 125 and Suman were moderately resistant,	(14)
	25	IC 8924 and IC 11312 were resistant; IC 8955, IC 9125, IC 10166, IC 10489 and IC 11379 as moderately resistant	(40)
	17	ML 62, ML 80, TT 8E, Pusa 103, PDM 14	(47)
	33	T 44 had minimum eggs of nematode	(23)
Blackgram	17	Ratnapur 1, UG 201 and UG 135	(47)

For rating the mungbean and urdbean varieties, scales have been developed on the basis of number of females penetrating the root of plant. One of them proposed by Chavada *et al.* is given below (14) :

Immune = No female detected penetrating the roots
Resistant = 1-10 females/plant
Moderately resistant = 11-20 females/plant
Susceptible = 21-30 females/plant
Highly susceptible = More than 30 females/plant

Control :

Chemical : Carbofuran incorporated in the soil 45 days after planting or before that @ 1 and 6 kg/ha (a.i.) respectively control the reniform nematode on mungbean (12). Soil fumigation by DBCP and metham sodium at 30 l/ha 15 days prior to sowing is reported to enhance the plant growth of blackgram and to reduce the multiplication of reniform nematode. The treatment costs about Rs 3000 to 3500 per hectare which may not be economic (33). The root penetration of pigeonpea is reduced greatly by spraying the plants with 2500 ppm of oxamyl (38). The control of reniform nematodes in mungbean and urdbean is reported by seed treatment with aldicarb, sulfon and carbofuran flowable @ 3 g a.i. w/w also and it is most economic (33).

Cultural : In well drained upland fields, 2 croppings of corn or sorghum or three clean fallows check the population of *R. reniformis* while three croppings of mungbean or soybean favour the population build up (12). Since the nematode has very wide host range, its control through crop rotation becomes difficult. *Coffea robusta, Canavalia ensiformis, Stizolobium deeringianum, Leucaena glauca, Eleusine coracana, Cynodona, Cynodona dactylon*, sugarcane, chillies, maize, and sorghum are resistant. Two successive crops of sorghum or maize, preceding mungbean, are reported effective in controlling the reniform nematode in Brazile (38). Rotation with sugarcane and the Pangola grass are recommended for its control in Puerto Rico (38). In Philippines, flooding the field or cropping flooded rice for 1-3 seasons reduced the nematode population significantly (11).

5.3. CYST NEMATODE :

Mungbean and urdbean were reported as the host of soybean cyst nematode in United States of America (21,38). In India, the pigeonpea cyst nematode has been reported to infect several host plants (leguminous) including mungbean and urdbean (28). These species of nematode occur commonly causing considerable damage to mungbean and urdbean crops.

Symptoms :

Lemon shaped, swollen females are found sticking with roots of the infected plant. When the female dies, its body wall becomes brown and hardened to a tough cyst which contains embryonated eggs. On the root surface of severely infected plant, the white swollen cysts are clearly visible. The field symptoms are the yellowing of the foliage; stunted growth; late emergence of leaves, flowers and foliage and ultimately poor yield. These symptoms usually occur in patches in the early stage of field infestation.

Causal organisms :

Heterodera glycines Ichinohe
Heterodera cajani Koshy

Heterodera glycines is reported to infect the mung bean in America (21) while *H.cajani* infects mungbean as well as urdbean in India (28). The cysts containing the eggs are easily detached from the host tissues and remain in the soil. The first moult occurs in egg and the second stage juveniles emerge from eggs which attach themselves with the host root just behind the root tip. After the fourth moult inside the tissues, the male larvae leave the host while females remain sedentary and attain almost the spherical shape. The adult female has its head attached to the root and the body projects outside it which later develops in to a pale to dark brown cyst containing eggs. The cyst remains in the soil for several years and protects the eggs against the adverse environmental conditions. The *H.cajani* reduces the plant growth of greengram and blackgram and the extent of reduction depends on the quantity of the initial inoculum incorporated in soil (51).

Control :

Resistance against the cyst nematodes has not yet been identified in mungbean and blackgram varieties except mungbean variety Jumbo reported to be resistant against *H.glycines* (47). As regards the cultural practices, *H.cajani* is confined to members of Leguminosae and thus may be controlled by crop rotation practice but *H.glycines* has a very wide host range. The resistant hosts like cotton & maize can be used in crop rotation for controlling the cyst nematodes.

Application of carbamate nematicides like aldicarb, oxamyl and carbofuran at the rate of 2-11 kg a.i./ha is effective. Temic has been reported to reduce the population of *H.cajani* in greengram and blackgram under the laboratory conditions (51).

5.4. ROOT LESION NEMATODE :

Pratylenchus spp. are commonly known as `root lesion' or `meadow' nematodes. They are the migratory endoparasites on several crops and cause considerable damage when present in adequate numbers. The nematode is found to occur on mungbean in India (54) and Philippines (20) and urdbean in India (34) under laboratory and/or natural conditions.

Symptoms :

The root lesion nematodes usually destroy the outer root tissues and develop extensive lesions when they are present in large number. The lesions are elongate, narrow and dark coloured. After some times they leave the older tissues and invade the newer suitable ones. Six or more nematodes /g soil reduce the germination in greengram variety Pusa Baisakhi and the seedling growth is adversely affected (54). The damage is increased due to the necrosis of root tissues which are invaded by the secondary organisms like *Fusarium solani* and *Macrophomina phaseolina*.

Causal organism :

Pratylenchus zeae Grahams
Pratylenchus mulchandi

Pratylenchus mulchandi was reported to infect blackgram under artificially inoculated conditions in India (34). The nematode was non specific in choosing the feeding sites and feed on root hairs, cortical or epidermal cells of rootlets and main root. The nematode multiplied very fast on callus of blackgram in the laboratory (35). Under natural conditions *Pratylenchus zeae* has been reported occurring on greengram in India (54). The reproduction in *Pratylenchus* spp. is bisexual. The eggs are deposited in root tissues. The first moult occurs within the egg producing the second stage juvenile. All the stages, specially adults and the pre-adults, move in to and out of roots and also move within the root.

Control :

Seed treatment by aldicarb 1 % a.i., carbofuran 0.5 % a.i. and phorate 1 % a.i. are effective in checking the root lesion nematode (53). They are not phytotoxic to greengram. Aldicarb or phorate at 1.5 % is reported to check mungbean root penetration by *P.zeae* up to 15 days (53).

5.5. OTHER NEMATODES :

Besides the above described nematodes, several other species have been reported associated with greengram and blackgram and they may cause damage under certain conditions. The information is summarized in Table 5.3.

Table 5.3

Minor nematodes found associated with mungbean and urdbean crops

Scientific name	Common name	Crop infected	Reference
Aphelenchoides sp.	Foliar nematode	Mungbean	(55)
Aphelenchus sp.		Mungbean	(55)
Criconemoides sp.	Ring nematode	Mungbean	(38)
Macroposthonia sp.	Ring nematode	Mungbean	(38)
Helicotylenchus sp.	Spiral nematode	Mungbean	(1,38)
Hemicriconemoides sp.	-	Mungbean	(38)
Hoplolaimus indicus	Lance nematode	Urdbean	(18)
Telotylenchus indicus	-	Mungbean	(38)
Tylenchorhynchus vulgaris	Stunt nematode	Mungbean	(38)
Tylenchus sp.	Surface nematode	Mung bean	(55)

In general,several methods including use of chemicals and soil amendments have been tried and found effective against mung bean nematodes. They are controlled best by soil amendments with wheat straw + neem seed cake + NPK or wheat straw + neem seed cake + 1/2 of NPK (44) in wheat followed by mungbean and maize cropping system. Neem cake is most effective followed by sesame and groundnut cakes for reducing the root galling and increasing the plant growth and yield of mungbean (33). Use of nematicides like DD, EDB, DBCP, aldicarb and dazomate is reported to influence the rhizosphere microbes of mungbean in addition of controlling neamatodes (49).

5.6 MITES :

The phytophagous mites are the important yield reducing agents attacking most of the crop plants. They belong mainly to the families Tetranychidae and Eriophyidae. An unidentified mite was observed assocaited with the greengram leaves at R.A.K. College of Agriculture, Sehore, India during the rainy season of 1990. The affected plants remained stunted with reduced or no flowering. The leaves were dark green in colour with smoky appearance on the upper surface. The interveinal areas on the lower surface of the leaves became pinkish brown in colour but the colour of veins and veinlets did not change. Affected leaves curled downward from their margins (Figure 5.2). The pod formation was curtailed greatly and those formed were not filled normally. Some of the upper leaves were deformed and did not grow normally.

REFERENCES

1. Ali, S.S. 1989. Plant parasitic nematodes associated with pulse crops. *National Symposium on New Frontiers in Pulses Research and Development, Kanpur*, India, 109 p.

2. Anonymous. 1984. Progress Report, All India Coordinated Research Project on Nematodes, IARI, New Delhi, 24 pp.

3. Anonymous. 1989. Consolidated Biennial Report of All India Coordinated Research Project on Plant Parasitic Nematodes with integrated approach for their control. *Biennial Workshop Bhubaneswar, April 24-27* : 35-37.

4. Bajet, N.B. and Castillo M.B. 1974. Effect of *Rotylenchulus reniformis* inoculations on mung bean, soybean and peanut. *Philippine Phytopathology* 10 : 50-55.

5. Bopaiah, B.M., Patil R.B. and Reddy D.D.R. 1976. Effect of *Meloidogyne javanica* on nodulation and symbiotic nitrogen fixation in mung, *Vigna radiata. Indian Jouranal of Nematology* 6 :124-130.

6. Castillo, M.B. 1971. Reniform nematode, *Rotylenchulus* sp. in mungo, soybean and peanut soils at the UPCA, Central Experiment Station. *Philippine Phytopathology* 7 :61-67.

7. Castillo, M.B. 1975. Plant parasitic nematodes associated with mung bean, soybean and peanut in the Philippines. *Philippine Agriculturist* 59 (3-4) : 91-99.

8. Castillo, M.B. 1976. Nematodes in cropping patterns. II. Control of *Meloidogyne incognita* through cropping patterns and cultural practices. *Phlippine Agriculturist* 59 : 295-312.

9. Castillo, M.B. 1977. Nematodes in cropping patterns. III. Composition and population of plant parasitic nematodes in selected cropping patterns in Balangas. *Philppine Agriculturist* 60 : 285-292.

10. Castillo, M.B., Alejar M.S. and Litsinger J.A. 1977. Pathogenic reaction and yield loss of mung bean to known population of *Rotylenchulus reniformis* and *Meloidogyne acrita. Philippine Agriculturist* 61 :12-24.

11. Castillo, M.B., and Litsinger J.A. 1977. Plant parasitic nematodes of mung bean in the Philippines. *First International Mungbean Symposium, AVRDC, Taiwan,* 195-200 pp.

12. Castillo, M.B., Bajet N.B. and Harwood R.R. 1976. Nematode in cropping patterns. I. population of *Rotylenchulus reniformis* on successive monocultured crops. *Philippine Agriculturist* 59 :288-294.

13. Catibog, C.S. and Castillo M.B. 1975. Pathogenicity of *Meloidogyne javanica* on mung bean (*Phaseolus aureus* Roxb.). *Philippine Agriculturist* 59 (5-6) : 189 -195.

14. Chavda, J.C., Patel B.A. and Patel, D.J. 1986. Susceptibility of certain mung bean varieties to reniform nematode, *Rotylenchulus reniformis*. *Pulse Crops Newsletter* 6: 52.

15. Chavda, J.C., Patel B.A. and Patel, D.J. 1988. Reaction of some green gram varieties to root knot nematodes. *Indian Journal of Nematology* 18 : 127.

16. Das, S.N., Mishra C.D. and Mohanty K.C. 1979. *Indian Journal of Nematology* 9 : 74.

17. Das, S.N., Routray B.N. and Sahoo H. 1986. Reaction of some green gram varieties to root knot nematodes, *Meloidogyne incognita*. *Indian Phytopathology* 39 : 126-128.

18. Das, S.N., Sahu H.S. and Ramana K.V. 1970. Host range and pathogenicity of *Hoplolaimus indicus* Sher. 1963 on sugarcane with some observations on its survival in fallow soil. *Indian Phytopathology* 23 : 157-158.

19. Datta, S. and Trivedi P.C. 1987. Changes in ascorbic acid concentration in root knot infected mung *(Vigna radiata)* and Guar *(Cyamopsis tetragonoloba)*. *Indian Journal of Nematology* 17 : 125-126.

20. Davide, R.G. 1972. Nematodes of Philippines crops and their control. *Philippine Phytopathology* 2 : 34.

21. Epps, J.M. and Chambers A.V. 1959. Mung bean *(Phaseolus aureus)* a host of soybean cyst nematode *(Heterodera glycine)*. *Plant Disease Reporter* 43 : 981-982.

22. Gupta, D.C. and Yadav B.S. 1979. Studies on the pathogenicity of the reniform nematode, *Rotylenchulus reniformis* to Urad, *Vigna mungo* (L.) Wilczek. *Indian Journal of Nematology* 9 : 48-50.

23. Gupta, D.C. and Yadav B.S. 1982. Note on the pathogenicity and the relative susceptibility of green gram varieties to *Rotylenchulus reniformis*. *Indian Journal of Nematology* 52 : 41-42.

24. Handa, D.K. and Mishra Alka. 1989. Assessment of losses and evaluation of mung bean genotypes against root knot nematodes. *Indian Journal of Pulses Research* 2 : 152-155.

25. Hussaini, S.S. and Seshadri A.R. 1975. Interrelationship between *Meloidogyne incognita* and *Rhizobium* sp. on mung bean. *Indian Journal of Nematology* 5 : 189-199.

26. Hussaini, S.S. and Seshadri A.R. 1976. Resistance in some mung bean (*Vigna radiata*) varieties and breeding lines to the root knot nematode, *Meloidogyne incognita, Indian Journal of Nematology* 6: 131-137.

27. Jain, M.K. and Rewari R.B. 1977. A report on the infestation of urd *(Phaseolus mungo)* crop by root knot nematode. *Indian Journal of Mycology and Plant Pathology* 7 : 94.

28. Koshy, P.K. and Swarup G. 1972. Susceptibility of plants to pigeonpea cyst nematodes, *Heterodera cajani. Indian Journal of Nematology* 2 : 1-6.

29. Krishna Murthy, G.V.G. and Elias N.A. 1968. Additions to host range of *Meloidogyne incognita* causing root knot on tobacco in Hunsur, Mysore District. *Indian Phytopathology* 21 : 351-353.

30. Madamba, C.P. 1974. The control of plant parasitic nematodes on mungo and other field legumes. *In R.G. Davide, Abstracts of Nematode Research Studies in Philippine National Science Development Board, Manila, Philippines,* 31p.

31. Mathur, R.L., Handa D.K. and Mathur B.N. 1973. Resistance in varieties of cowpea, green gram and guar to root knot nematodes., *Meloidogyne javanica* and *Meloidogyne incognita. Indian Jouranl of Mycology and Plant Pathology* 3: 182-183.

32. Midha, R.L. and Yadav B.S. 1989. Control of reniform nematode through nematicides on black gram. *Indian Phytopathology* 42 : 295.

33. Mishra, S.D. and,Prasad S.K. 1974. Effect of soil ammendments on nematodes and crop yields. I. Oilseed cakes, organic matter, inorganic fertilizers and growth regulator on nematodes associated with wheat and their residual effect on mung.*Indian Journal of Nematology* 4: 1-19.

34. Nandkumar, C., Khera S. and Bhatnagar G.C. 1969. *In vitro* studies on the host range of *Pratylenchus mulchandi. Indian Phytopathology* 22 : 521-522.

35. Nandkumar, C. and Khera S. 1969. Propagation of *Pratylenchus mulchandi* on alfalfa and black gram callus tissues. *Indian Phytopathology* 22 : 458-461.

36. Nanjappa, C.K., Panda M. and Seshadri A.R. 1978. Effect of *Meloidogyne incognita and Rotylenchulus reniformis* on the emergence and survival of seedlings. *Third International Congress of Plant Pathology, Germany, Aug. 16-23.*

37. Panda, M. and Seshadri A.R. 1979. Effect of initial inoculum level of *Rotylenchulus reniformis* on different growth attributes of cowpea, mung bean and urid. *Indian Journal of Nematology* 9 : 59-60.

38. PANS. 1981. Pest Control in Tropical Grain Legumes. *PANS Manual No. 5*, 206 pp.

39. Pant, Veena, Hakim Sheila, Singh S.P. and Saxena S.K. 1983. Morphometrics of root knot nematodes, *Meloidogyne incognita* as influenced by different pulses. *Indian Journal of Mycology and Plant Pathology* 13 : 313-318.

40. Patel, B.A. chavda J.C. and Patel D.J. 1989. Screening of certain green gram lines against reniform nematode, *Rotylenchulus reniformis. International Nematology Network Newsletter* 6(2) : 25-26.

41. Patel, S.T., Patel H.V., Patel, M.C. and Patel D.J. 1986. Evaluation of certain promising mung bean lines against root knot nematode. *Pulse Crops Newsletter* 6 : 53.

42. Poehlman, J.M. 1977. What we have learned from the International mung bean nursaries. *First International Mung bean Symposium, AVRDC, Taiwan,* 97-100 pp.

43. Prasad, S.K., Chawala M.L., Kumar S. and Saxena H.P. 1971. Root knot nematode, `*Meloidogyne javanica* (Troub 1885] Chitwood and stem borer, *Stomopteryx nertaria* Meryrick affecting green gram, *Phaseolus aureus. Indian Journal of Entomology* 33 : 55-60.

44. Prasad, S.K., Mishra S.D. and Gaur A.C. 1972. Effect of soil amendments on nematodes associated with wheat followed by mung and maize. *Indian Journal of Entomology* 34 : 307-311.

45. Rachie, K.O. 1974. Grain legumes of the lowland tropics : Mung bean. In K.O. Rachie and L.M. Roberts. *Advances in Agronomy* 26 : 62-67.

46. Raut, S.P. 1980. Effect of initial inoculum levels of *Meloidogyne incognita* on plant growth and Rhizobial nodulation of mung bean *Indian Phytopathology* 33 : 351-353.

47. Routray, B.N., Sahoo H. and Das S.N. 1986. Evaluation of green gram and black gram varieties against reniform nematode, *Rotylenchulus reniformis. Indian Jouranl of Nematology* 16 : 27-29.

48. Singh, H.K. and Ranganaikulu Y.S. 1989. Screening of mung bean cultivars for their resistance against root knot nematode, *Meloidogyne incognita. National Symposium New Frontiers in Pulses Research and Development, Kanpur,* India, 107 p.

49. Singh, I. and Prasad S.K. 1978.Influence of pesticides on rhizpsphere microbial population of mung and wheat crops. *Indian Journal of Nematology* 8 : 102-109.

50. Singh, R.N. 1972. Root knot disease of urd (*Phaseolus mungo*) and mung (*Phaseolus aureus*) in India. *Indian Journal of Mycology and Plant Pathology* 2 : 87.

51. Sobita Devi, L.R. and Gupta Pramila. 1988. Pathogenicity and management of *Heterodera cajani* on some pulse crops. *Indian Phytopathology* 4 : 470-472.

52. Toledo, R.R. and davide R.G. 1969. Reactions of different vegetables and field crops to infection by *Meloidogyne javanica. Philippine Phytopathology* 5 : 45-51.

53. Walia, R.K. and Seshadri A.R. 1986. Chemical control of root lesion nematodes on green gram by seed coating methods. *Indian Journal of Agricultural Sciences* 56 : 735-737.

54. Walia, R.K. and Seshadri A.R. 1986. Pathogenicity of root lesion nematode to green gram. *Indian Journal of Agricultural Sciences* 56 : 792-795.

55. Yang. C.Y. 1977. Mungbean diseases and control. *First International Mungbean symposium, AVRDC, Taiwan,* 141-146 pp.

56. Yein, B.R., Singh H. and Chhabra H.K. 1977. Effect of pesticides on the root knot nematode infesting mung. *Indian Journal of Nematology* 7 : 117-122.

CHAPTER 6
SEED PATHOLOGY

Seed may contain large numer of microorganisms which may prove hazardous for the seed itself or the new plant created from it. The associated microorganism may be pathogenic, weak parasite or a saprophyte. It may be associated internally or externally with the seed or as concomitant contamination. It is mentioned that greengram and blackgram carry more microogranisms externally than internally (60). In fact, seed is the vehicle for conveying serious pathogens to new area. If seed of excellent germination and purity is loaded with serious pathogens, usually undetectable by the naked eye, the crop grown from it and even the neighbouring crop may be destroyed. Soil may become contaminated from the infected seed and the subsequent crops will suffer from the diseases thus established. Seeds of blackgram as well as greengram are reported to harbour large number of fungi, bacteria and viruses responsible for causing many serious diseases which substantially reduce one of the worlds vital source of protein.

6.1 FUNGI :

6.1.1. Type of fungi associated :

i. **Field fungi** : They invade the seed while in the field on plant itself. Most common genera belonging to this group are *Alternaria, Drechslera, Fusarium, Diplodia, Chaetomium, Phoma, Colletotrichum, Rhizoctonia* etc. The field fungi may cause discolouration of grain, weakening or death of embryo and variety of other diseases. Many of these fungi are pathogenic to the crop and cause diseases at later stages of plant growth (70, 76, 77). They may cause seed rot, seedling blight, root rot, collar rot, hypocotyl invasion, foliage spotting /blight, pod rot/blight etc. (Tables 6.1 & 6.2). Most of the field fungi associated with greengram and blackgram seeds persist as long as 120 days in storage at RH 45-78 % and 6.2 - 29.6° C temperature. They are replaced by storage fungi between 120-300 days at 28.2-87 % RH (77).

ii. **Storage fungi** : These are the fungi which invade the grain rendering unsuitable for use as seed and harmful for human beings, birds and animals due to production of certain toxins etc. They increase during the storage period.

6.1.2. Fungi associated with greengram and blackgram :

Large number of fungi are associated with seed of these two important pulse crops (Table 6.1). Hussain and Ahmad (38) isolated 36 species of fungi from seeds of greengram and blackgram including some other pulses. Of these 12 were the species of *Aspergillus* only. Similarly Gupta and Gupta (35, 37) isolated 34 fungi from stored seeds of greengram alone. The significance of the storage fungi of blackgram and greengram has been discussed in detail (78).

6.1.3. Effect of fungi on seed quality :

The fungi reduce the seed quality in three ways :

i. **Adverse effect on viability, weight and colour** : The fungi associated with seed reduce its viability resulting in seed rot, seedling necrosis and reduced vigour of seedling (Fig. 6.1). They also discolour the seed and reduce its weight by chemical degradation (17, 78). The increased incidence of moulds on greengram seeds during the storage, increases the loss in seed germinability (37). These adverse effects of fungi on seed are due to certain metabolites, particularly enzymes and toxins produced by them during their growth. The culture filtrates of *Fusarium oxysporum, Aspergillus flavus, Alternaria alternata, Curvularia lunata* and *Drechslera halodes* are reported to reduce the germination and root shoot elongation of blackgram (47, 48). In case of *F. oxysporum*, the inhibitory effect increases with the age of the culture (26). The culture filtrates of *A.flavus, Aspergillus niger* and *Penicillium* sp. cause phytotoxic effect on greengram resulting in reduced seed germination and poor seedling growth (89). On the other hand, extracts of *Chaetomium globosum, Fusarium moniliforme, Curvularia pallescens, Botryodiplodia theobromae* and *Macrophomina phaseolina* reduce the seed germination considerably but do not affect the seedling growth much. The extract of *Alternaria tenuis* is reported to incite the albinism in seedlings of greengram (71). Most of the seed borne fungi discolour the seed and cause shrivelling and reduction in their size (15).

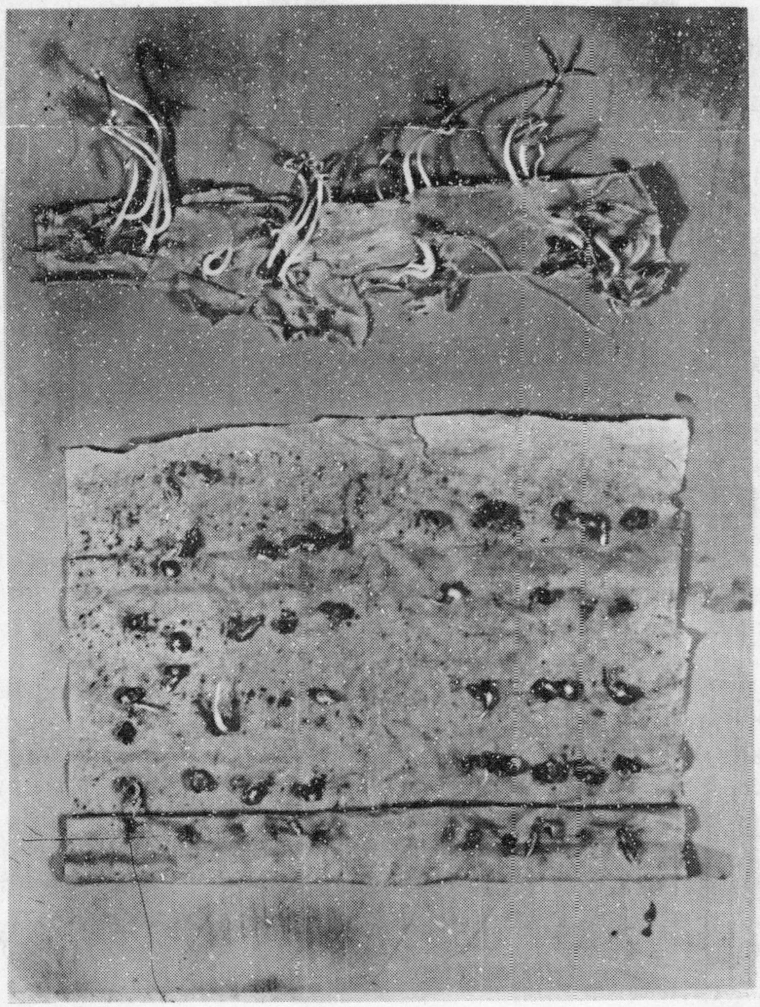

Fig.6.1.
Seedlings from healthy black gram seed (above). Mortality by seed-borne *Rhizoctonia batatícola* (below).

ii. **Seed becomes carrier of diseases** : Most of the diseases of greengram and blackgram are seed transmitted. The inoculum of the pathogen reaches in the new area through the infected seed and produces seedling and foliar diseases (Table 6.2). The seed borne fungi cause inhibition of seed germination, seed rot, radicle necrosis, seedling necrosis and damping off of seedlings in greengram as well as blackgram (76). Many of the seed borne fungi cause root rot, stem rot, wilt, leaf spots, foliage blight and pod diseases in both the crops (49, 57, 69, 74, 75). These fungi are listed in the table and are very much important with the view point of quarantine regulations (49).

iii. **Indirect effects** : Some of the seed borne fungi affect the growth of plant indirectly. *Phoma exigua, Rhizoctonia solani, F.oxysporum, Rhizoctonia bataticola* and *Sclerotium rolfsii* are reported to inhibit the rhizobial nodulation completely (28).

6.1.4. Effect of fungi on food value :

The seed fungi, by their biological activity, consume some of the seed contents and produce some of the newer ones altering their food value greatly.

i. **Effect on nutritive value** : Pulses are the good source of protein for the human diet. The fungi cause their degradation during the storage. High freqency of *A.flavus* associated with greengram seeds reduces the sugar contents of seed (15). *C. lunata* also causes depletion of sugar in greengram seeds (74). It has been noted that the fungal infection decreases the starch but increases reducing sugars which is due to break down of complex sugars in to simpler ones by the activity of ß-amylase (16). *F. oxysporum* increases reducing sugar contents of mungbean seed to a greater extent (31).

The seed protein is also depleted by fungal infection in greengram as well as blackgram seeds (16, 24, 31, 35, 36, 50). It is reported that the decrease in starch and protein is more at 75-96 % RH while at lower RH these changes are insignificant (90). In case of urdbean seed, the fungal infection reduces the protein contents but increases soluble nitrogen which is attributed to the hydrolysis of the protein by fungi (42). The maximum change in the protein of blackgram seed occurs at 96 % RH which corresponds to 20.4 % seed moisture (88). The corresponding seed moisture levels at 33,

55 and 75 % RH are 11.2, 13.4 and 16.8 % respectively in case of blackgram. The protein reduction is positively correlated with it (88). The amino acids are also affected by the fungal infection but no definite pattern in their quantity or quality is expressed.

The infection of seed by *A.flavus* or other fungi increases some other acids particularly fumaric acid and tartaric acid and produces oxalic acid (15,16) while succinic acid is reduced. It also increases fatty acids in greengram (24). The artificial inoculation of greengram seeds with heavy inoculum/spore suspension of *R.solani, Phoma exigua* and *Colletotrichum capsici* reduces the ascorbic acid contents of germinating seeds (21).

ii. **Mycotoxin production** : During the storage period, certain fungi produce toxins which may be hazardous to the health of consumers. Among such toxins, aflatoxin produced by *A.flavus* is important and causes hepatotoxic and hepatocarcinogenic effects. Gupta *et al.* (36) isolated 16 strains of *A.flavus* from 78 seed and '*Dal*' samples of greengram collected in different seasons from the market of Agra, India. When tested by the chlorosis index method, 12 strains were found aflatoxigenic. The strains having deeper yellow to orange colour, wrinkled growth and larger sclerotial production, favoured higher chlorosis which might be related to aflatoxin production. The fungus *A.flavus* group dominated in 40 % samples of green gram seed collected from Bhagalpur, India (66,67). Fifteen samples showed aflatoxin concentrations up to 2800 ppb. Out of 124 isolates of *A.flavus*, 46 produced aflatoxin. Only one isolate produced all the three aflatoxins i.e. B_1, B_2 & G_1. The elaboration of aflatoxin B_1 ranged between traces to 8400 ppb.

In addition to hazardous effect of aflatoxin to human and animal health, it affects germinability of seed also (97). The aflatoxin extracted from *A.flavus* on rice is reported to reduce the germination and allantoinase activity of germinating seeds of greengram and blackgram (97).

6.1.5. Deterioration of seed by fungi :

Fungi are well known to produce certain enzymes which break down complex ingredients of seed in to other compounds ultimately affecting its viability and the nutritive value The infection of greengram seeds by *F.oxysporum* results in maceration of seed tissues due to cellulolytic and pectinolytic enzymes produced by the fungus (31). The fungi *P.exigua*, and *Graphium penicillioides* are reported to produce the hydrolytic enzymes (cellulolytic, pectinolytic, amylolytic and proteolytic) for colonizing greengram seeds (20). Cellulolytic are the cell wall degrading enzymes and hence, play an important role in the pathogenicity. They are produced by *P. exigua* isolated from the greengram seed (27). It also produces pectin lyase and pectic acid lyase *in vitro* (19). *F. oxysporum* and *P. exigua* increase the production of alkaline as well as acidic phosphatases by germinating greengram seeds (25).

The production of enzymes by seed borne fungi is influenced by several factors. Fungicides (antracol and panoctine), phenols (catechol and gallic acid) and growth substances (GA and Yeast extract) retard the production of pectin lyase and pectic acid lyase by *P. exigua* (19). Catechol, salicylic acid, pyrogallic acid and calixin inhibit cellulase production by *P. exigua* and the inhibitory effect increases with the concentration of the chemical. In the same way secretion of pectolytic enzymes by *P.exigua* is influenced by different chemicals (29). The secretion of PME is reduced by antracol and panoctine. Yeast extract and GA inhibit the secretion of endo-PG but they enhance secretion of exo-PMG and endo-PMG respectively. Catechol inhibits endo-PAL, antracol and GA inhibit exo-PAL while phenols and fungicides inhibit endo and exo-PAL production by *P.exigua*.

Besides the secretion of enzymes and toxins, the fungal infection increases the moisture content of seed and the temperature of the store which accelerate the loss in viability of seed and the biochemical changes reducing the nutritive value of seed and ultimately making it unfit for consumption.

6.1.6. Favourable effects of fungi :

The seed fungi also produce certain growth promoting substances like auxins or antifungal chemicals which may show beneficial effects under certain conditions. A substantial increase in germination of greengram seeds has been observed when treated with the culture filtrates of *A. alternata, C. lunata* and *Helminthosporium oryzae* (89). In general, the culture filtrates are reported to reduce the fungal incidence on seed probably due to the presence of antonistic metabolites (89). The fungus *P. exigua* isolated from greengram seeds reduces the nodulation under sterilized soil conditions but if the seed is incolulated simultaneously with *Rhizobium* also then the root nodulation is increased in greengram (28).

6.1.7. Effect of seed leachate on seed fungi :

During the process of germination, the seed exudes certain chemicals which may directly influence the growth of a fungus associated with it. The leachates from seeds of greengram and blackgram are reported to decrease the spore germination of *Fusarium solani, F. oxysporum* and *C. lunata* (12). The germinating seeds of greengram produce the volatile exudates which affect the spore germination of *A. alternata, C. lunata, F. oxysporum* and *Drechslera rostrata* (18). However, it has been mentioned that the seed leachates from blackgram are able to inhibit the growth of externally associated fungi and not the internally associated ones (72). A very interesting observation has been made by Bhargava and Shukla (12) that the seed coat leachates of greengram and blackgram can decrease the spore germination of certain fungi but the culture filtrates of these fungi are able to inhibit the seed germination. It indicates that in association the dominance will be of the party which becomes active first. The effect of leachates from greengram seed may be inhibitory or stimulatory to spore germination and it is dependent on the fungus species tested, duration of soaking the seed and the crop (73).

The greengram seed leachate contains atleast 6 sugars, 12 amino acids and 5 organic acids (73). The amino acids may also alter the toxicity of seed treating fungicide to a greater extent. Jhooty and Bains detected 15 amino acids in the leachates of greengram seeds of which 7 decreased the toxicity of benomyl against *F. solani* and *Macrophomina phaseolina* while 2 increased it considerably (44).

6.1.8. Methods of detection :

For detecting the seed borne fungi, the "Standard Blotter Method" is recommended (39) but efforts have been made to find the most specific method for detecting the specific pathogen by adopting certain modifications. The fungi *A. tenuis, Rhizoctonia bataticola, Cladosporium fulvum* and *Fusarium* sp. were detected by all the three methods i.e. `Agar Plate Method', Standard Blotter

Method' and 'Moist Sand Technique' (93) by using surface sterilized as well as unsterilized seeds of greengram suggesting their deep seated infections. Suhag and Suryanarayana used several methods for detecting seed borne funqi of blackgram and lentil (94).

The 'Agar Plate Method' and 'Standard Blotter Method' as such or modified by soaking blotters with 2, 4-D, have been found suitable for detecting *Fusarium moniliforme* and *F. semitectum* associated with greengram seeds. (2, 31). The modified blotter method with water at 6 pH and modified Czapek's agar have been found selective for detection of *Fusarium* spp (2, 31). Out of seven methods compared for detecting *Fusarium* spp. in greengram seed, 'Agar Plate Method' was best for detecting *F. moniliforme* and *F. semitectum* (6). The 'Standard Blotter Method' with modification of soaking the blotters in 2, 4-D solution was suitable to all the three species of *Fusarium* of greengram.

Out of several methods used for detecting the *Colletotrichum* spp. on blackgram seeds, 'Standard Blotter Method' modified by using 2, 4-D solution for blotter soaking was best. The modified blotter method using water of 6 pH and modified 'Deep Freezing Method' in which blotters were saturated with Agrimycin or streptocycline solution also gave higher counts of *Colletotrichum* spp. in blackgram (87).

The isolation of blackgram seed fungi is reported to be influenced greatly by the medium, its pH and the temperature of incubation. Rose bengal agar favours the isolation of *M. phaseolina* while D-5 is suitable for '*Fusarium* (13).

6.1.9. Control

i. **Seed treatment** : Large number of fungicides have been tested and found effective against seed borne fungi of greengram and blackgram. Thiram is one of the old recommendation for treating the seeds of pulse crops. It is the most effective fungicide against large number of fungi of both the crops when used alone (50, 80, 93) or in combination with other fungicides like PCNB (33) and carbendazim. Other fungicides found effective against seed borne pathogens of greengram and blackgram are vitavax, demosan, benlate (40), ceresan, captan (93, 98), mancozeb (100) and carbendazim (1, 31, 100). These fungicides disinfect the seed and improve the stand of plants in the field increasing the yield of mungbean as well as urdbean. However, certain considerations are important to be kept in mind before recommending a pre-storage seed treating fungicide to protect it from the spoilage by fungi:

a. The fungicide should be effective against maximum number of potentially damaging seed borne pathogens.
b. It should not be injurious to seed during the storage.
c. It should not be toxic to *Rhizobium*.
d. It should be cheap, easily available and with suitably long residual effects.

Thiram and captan have been reported to be effective seed dressers in maintaining the yield of blackgram after 5 months or more in the storage (98). In case of greengram, improved seed germination is reported by pre-storage seed treatment with bavistin, thiram or dithane M-45 (100). Charya and Reddy (30) worked out the efficacy of several fungicides against seed borne fungi of greengram and discussed the results in relation to phytotoxicity and human hazards. The organomercurials are expected to be hazardous for human health and injurious to seed as well as *Rhizobium*. In the field trials, no inhibitory impact of fungicides is observed on the survival of *Rhizobium* of mungbean when used at the recommended doses but higher doses of fungicides adversely affect nodulation and growth parameters (41). Seed treatment with Agrimycin 250 ppm is reported to increase the nodulation and grain weight.

Based on the results of several experiments conducted on seed treatment of greengram and blackgram, thiram (0.25 %), carbendazim (0.1 %), mancozeb (0.25 %), captan (0.25 %) or thiram + carbendazim (2:1) @ 2.5 g/Kg of seed are the promising seed dressing fungicides for greengram and blackgram. Carbendazim is reported to increase the plant stand and yield considerably (1, 41).

ii. **Fumigation** : The fungicidal seed treatment can not be practiced when the seed is stored for consumption purpose. In such cases fumigation is the only alternative to control the damage by seed mycoflora. Out of 19 volatile compounds tested, only acetone, aniline and amyl alcohol could effectively control the greengram grain spoilage due to *R.solani* during post harvest storage (23). In case of blackgram, ethylene dibromide is more effective followed by phosphine. However, they have little or no effect on *A.terreus*,

F.oxysporum, Drechslera turcica and *Mucor* sp. (58). In urdbean, the ethylene dichloride is more effective against bacteria, ethyline dibromide against fungi and phosphine agianst Actinomycetes (50). These fumigants also reduce the free fatty acidity and reducing sugar contents of the seed. EDC and EDB affect the viability of seeds but phosphine has no adverse effect on the germination. The optimum dose of the fumigants under laboratory conditions is reported 100 mg/l. An essential oil present in seeds of mustard (allyl-isothiocyanate) is reported to be more effective as preservative even at 0.1 % against fungal infection of seeds of several crops including blackgram. It is better than the propionic acid and has no phytotoxic effect (54).

iii. **Proper storage** : The storage conditions specially temperature, RH, seed moisture and aeration play an important role in causing infection of seeds and their deterioration by fungi in the store (22). The greengram seeds can be damaged by fungi only when they contain more than 11 % seed moisture (61). The proper seed moisture can be obtained by timely harvesting and proper drying of seeds before storage. Early harvesting results in high seed moisture while harvesting delayed after the field maturity of greengram reduces the yield as well as seed quality by weathering. The later harvested seeds have higher incidence of infection by seed borne fungi specially *Fusarium* spp. (53). The infection increases after onset of rainy season. During the storage, RH controls the seed moisture as they are hygroscopic in nature. In case of blackgram, the corresponding seed moisture levels at 33, 55, 75 and 96 % RH are 11.2, 13.4, 16.8 and 20.4 % respectively (88). The seed moisture of greengram at 96 % RH will be about 21 % (61). In a test, the infection of greengram by *A.flavus, A.fumigatus* and *A.parasiticus* became more than 80 % after 18 week storage at 95 % RH while in case of *A.niger* it was 90 % within 9 weeks under the same conditions. At 30 and 50 % RH the infection by fungi was minimum and the seed did not show any decline in germination even after 18 weeks in storage (99). Higher temperature accelerate the deterioration of seed by fungi. The seed fungi are influenced by different storage conditions to such an extent that their quality as well as quantity differ from store to store (13).

6.2. BACTERIA :

It is reported that greengram seeds contain more bacteria than fungi which includes mostly gram negative rods (60). The bacterial leaf spot pathogen of greengram as well as blackgram are seed borne. The seed infection by *Xanthomonas campestris* pv *vignaeradiatae* is easily detected by growing on test during monsoon but not during summer (91). The wounded seed growing on test is reported to facilitate the detection of minimum level of seed inoculum but not the per cent seed infection by *X.campestris* pv. *vignaeradiatae*. The seed borne inoculum can be eradicated by hot water treatment at 52°C for 20 minutes or seed soaking in 700 ppm agrimycin (34). Similar results have been obtained by Thind *et al.* (96). Agrimycin is not inhibitory to the *Rhizobium* but it increases nodulation in greengram plant (41).

6.3. VIRUSES :

Several viruses have been reported to be seed transmitted in greengram as well as blackgram (Table 6.3). Out of these, some cause important diseases. The Beam Common Mosaic Virus (BCMV) is easily detected in seed by indicator inoculation, growing on and microprecipitin tests (4) and it has been located in seed coat, cotyledons and primary axis of the seed (5). Urdbean Leaf Crinkle Virus (ULCV) is another important seed transmitted virus reported from India (10). The infected seeds are characterized by shrivelling and brown discolouration of seeds and presence of oversize seeds in the lot (7). These characters can be used for identifying the infected seed lots (8) and the infection can be further confirmed by growing on and indicator inoculation tests. The virus is detected in seed coat, cotyledons and primary axis of urdbean seeds (9). Sharma and Dubey (81) reported aphid transmitted ULCV confined to cotyledons and embryo only. The ULCV can be eradicated from the blackgram seeds by hot water treatment at 60° C for 10-20 minutes or dry heat at 70°C for 10-30 minutes (11). However, it is reported that none of the chemicals and dry heat are able to eradicate the ULCV without affecting the seed viability but seed dip in 2-thiouracil or 8-azaguanine (0.05-0.1 %) for 30 minutes is effective (45). The wet heat at 65°C for 30 minutes or the solar heat treatment are also effective (45). The seeds of blackgram can be inoculated artificially by shaking them with the inoculum of ULCV for 30 seconds (33). The Blackgram Leaf Crinkle Virus (BLCV) studied by Narayanaswamy and Jaganathan is also seed borne and detected in plumule, radicle and cotyledons of blackgram but not in the seed coat (56).

Several methods have been used for detecting the seed transmitted viruses of greengram and blackgram and it has been suggested to use combination of methods for different viruses in plant quarantine laboratories (63,68).

Table 6.1
Fungi detected on greengram and blackgram seeds and their significance

Fungus	Significance		Reference
	Greengram	Blackgram	
Achaetomium strumarium	SR, SN	-	(95)
Alternaria sp.	UN	-	(43, 80)
A. longissima	UN	UN	(2)
A.porri	-	UN	(2)
A. tenuis	SD, GI, SN	SD, SN	(69, 70, 93)
Ambliosporium sp.	NP	-	(78)
Aspergillus sp.	SR	-	(32, 43)
Aspergillus avenaceus	NP	GI, SR	(78)
A. Candidus	GI	NP	(78)
A.flavus	SD, GI, MT	SD	(15, 36, 43, 85, 89)
A. flavipes	GI, SR, RN	-	(78)
A. fumigatus	GI, SR	SR, RN	(78, 99)
A.glaucus	-	UN	(58)
A.nidulans	SR	-	(78)
A. niger	GI	GI	(32,43,86,89)
A. ochracious	NP	NP	(43, 78)
A. parasiticus	UN	-	(99)
A. quadrilineatus	UN	GI, SR	(78)
A. sulphureus	UN	NP	(78)
A. stellatus	NP	GI, SR	(78)
A. sydowi	GI	UN	(78)
A. tamarii	GI, SR,RN	GI, SR	(78)
A. terreus	UN	NP	(58, 78)
A.variecolor	SR	UN	(78)
Cephalosporium sp.	-	UN	(2)
Chaetomium sp.	UN	UN	(2)
Chaetomium erresticum	GI	-	(78)
C. globosum	GI, SR	GI, SR	(78, 89)
C. indicum	GI	NP	(78)
C. olivaceum	UN	UN	(78)
C. spirale	GI	NP	(78)
Choanephora sp.	GI, SR	GI	(57, 78)
Cladosporium epiphyllum	-	GI, SR	(78)

(Contd...)

C. fulvum	SR, RN	-	(93)
C. tenuissium	UN	-	(32)
Cunninghamella echinulata NP	NP	*NP*	*(78)*
Curvularia geniculata	-	UN	(2)
C. lunata	GI, SR	UN	(2,32,84)
C.pallescens	GI	UN	(2,89)
C.verruculosa	UN	UN	(2)
Cylindrocladium scoparium	*UN*	-	*(52)*
Drechslera hawaiiensis	*SR, RN*	-	*(32)*
D.rostrata	UN	-	(57)
D.sativa	UN	UN	(2)
D. tetramera	UN	UN	(2)
D.turcica	-	UN	(58)
Epicoccum sp.	-	NP	(78)
Epicoccum purpurascens	AG	-	
Fusarium longipes	SD	UN	(2)
F. sporotrichioides	SR,SN	-	(62)
Diporthe phaseolorum	GI	-	(57,51'
Graphium penicillioides	*SD*	-	*(20)*
Helminthosporium sp.	UN	-	(43)
H. hawaiiensis	UN	UN	(15)
H. oryzae	UN	-	(89)
Heterosporium sp.	-	GI,SR,RN,SN,DO	*(78)*
Hormiscium sp.	GI,SR,RN,SN	GI,SR,RN,SN	(78)
Humicola sp.	NP	NP	(78)
Isariopsis sp.	GI,SR,RN	GI,SR	(78)
Melanospora sp.	UN	-	(2)
Memnoniella echinata	GI,SR,RN,SN	GI,SR	(78)
Monilia sp.	NP	NP	(78)
Mucor sp.	NP	UN	(78)
M.hiemalis	NP	NP	(78)
Nematospora coryli	Yeast spot	-	(55)
N.gossypii	Yeast spot	-	(55)
Penicillium chrysogenum UN	*UN*	-	*(85)*
P.crustcsum	GI,SR	GI,SR	(78,86)
P.cyclopium	UN	-	(78)
P. frequentans	GI	GI,SR	(78)
P.oxalicum	UN	GI,SR	(78)
P. wertmanni	GI,SR	GI,SR	(78)
Paecilomyces varioti	GI,SR,	GI,SR	(78)
Papulospora sp.	UN	GI,SR,RN,SN	(78)
Periconia sp.	NP	UN	(2,78)
Pestalotia sp.	UN	-	(43)
Phoma sp.	UN	UN	(2,51)
Phoma exigua	GI,SR	-	(17)
P.glomerata	-	GI	(86)
Pleospora infectoria	SR,RN	-	(32)
Rhizopus arrhizus	UN	-	(85)
R.nigricans	NP	NP	(78,93)

(Contd...)

Sporotrichum sp.	UN	-	(85)
Stachybotrys atra	GI,SR,RN	GI,SR,RN	(78)
Stemphylium sp.	-	UN	(2)
Trichoderma viride	-	AG	(78)
Trichothecium sp.	UN	AG	(78)
Trichurus sp.	GI	GI,SR,RN	(78)
Torula sp.	UN	UN	(78)
Ulocladium sp.	-	UN	(2)
Verticilium cinnabarinum	-	UN	(2)

Note:
AG : antagonistic
DO : damping off
GI : germination inhibition
MT : mycotoxin production
NP : non pathogenic
RN : radicle necrosis
SD : seed deterioration
SN : seedling necrosis
SR : seed rot
UN : unknown.

Table 6.2.
Seed borne fungi and bacteria causing diseases at later stage of crop growth

Pathogen	Disease symptoms		Reference
	Greengram	Blackgram	
Ascochyta sp.	LS	LS	(51)
Ascotricha chartarum	CR,LS	CR,LS	(74,75)
Botryodiplodia palmarum	SR,SB	-	(57)
B. theobromae	PR	PR	(51)
Cercospora kikuchii	LS	LS	(2,57)
C. canescens	LS	LS	(76)
C. cruenta	LS	LS	(76)
Colletotrichum dematium	SR,SB,LS	SR,SB,LS	(51,70)
C. graminearum	SR,SB	SR,SB	(51)
C. graminicola	-	SR,SB,LS,PB	(87)
C. lindemuthianum	SR,SB,LS,AN	SR,SB,LS,AN	(51,87)
C. truncatum	LS	SR,SB	(2,57,69,74,
Corynespora cassiicola	LS	-	(57)
Fusarium equiseti	SR,SB	LS,PI	(32,51,57)
F. moniliforme	GI	RR	(2,50,57)
F. oxysporum	SR,SB,PI,W	SR,SB,PI,W	(31,32,75,8: 83)
F. semitectum	DO, W	DO,W	(2,31,57,74)

(Contd.

F.solani	GI,RR	GI,RR	(2,31,57,84)
Macrophomina phaseolina	SR,SB,LS,PI	SR,SB,LS,PI	(32,40,57,1)
Myrothecium roridum	CR,LS	LS	(2,57)
Phoma exigua	SR,SD	LS	(24)
Phyllosticta phaseolina	-	LS	-
Rhizoctonia solani	SR,SD,WB	SR,SD,WB	(24) & others
Xanthomonas campestris pv vignaeradiatae	LS	-	(91,92)
Pseudomonas syringae pv phaseolicola	HB(Halo blight)	-	(Schmitthenner 1971)
Xanthomonas campestris pv phaseoli blackgram strain	LS	LS	(Kore and Shirshikar 1980)

Note : LS : leaf spot, CR : collar rot, PR : pod rot, SR : *seed* rot, SB : seedling blight, PB : pod blight, AN : anthracnose, PI : pod infection, GI : germination inhibition, RR : root rot, W : wilt, SD : seed deterioration, DO : damping off, WB : web blight.

Table 6.3
Viruses detected in greengram and blackgram seeds

Virus	Crop	Brief symptoms	Reference
Alfalfa Mosaic Virus	Greengram	Yellow mosaic	(Kaiser *et.al.* 1969)
Bean Commn Mosaic	Greengram blackgram	Mosaic mottle	(3,46)
Bean Yellow Mosaic	Greengram blackgram	Mosaic symptoms	(PANS 1981)
Blackgram Leaf Crinkle	Blackgram	Leaf crinkling	(56)
Black eye Cowpea Mosaic	Greengram - & blackgram	-	(65)
Broadbean Mottle virus (blackgram strain)	Blackgram	-	(63)
Cowpea aphid borne masaic	Grangram blackgram	Mosaic	(PANS, 1981)
Cowpea Yellow Mosaic	Greengram & black-gram	Bright yellow or light green mosaic	(PANS 1981)
Cucumber Mosaic Virus (mungbean strain)	Greengram	Yellow mosaic	(63)
Mungbean Mosaic Virus	Greengram	Mosaic	(63)
Southern bean mosaic	Greengram	Mosaic & mottling	(PANS, 1981)
Urdbean Leaf Crinkle Virus (ULCV)	Greengram & blackgram	Leaf crinkling	(10)

REFERENCES

1. Agrawal, S.C. and Nema Sushma, 1989. Effect of carbendazim on Macrophomina leaf blight of blackgram and greengram. *Indian Journal of Plant Protection* 17 : 147-149.

2. Agrawal, V.K., Mathur S.B. and Neergaard P. 1972. Some aspects of seed health testing with respect to seed borne fungi of rice, wheat, blackgram, greengram and soybean grown in Inda. *Indian Phytopathology* 25 : 91-100.

3. Agrawal, V.K.,Nene Y.L. and Beniwal S.P.S. 1976. Influence of bean common mosaic virus infection on the flower organelles, seed characters and yield of urid bean. *Indian Phytopathology* 29 : 444-445.

4. Agrawal, V.K., Nene Y.L. and Beniwal S.P.S. 1977. Detection of bean common mosaic virus in urid bean (*Phaseolus mungo*) seeds. *Seed Science & Technology*, 5 : 619-625.

5. Agrawal, V.K., Nene Y.L. and Beniwal S.P.S. 1979. Location of bean common mosaic virus in urdbean seed. *Seed Science & Technology* 7 : 455-458.

6. Baghel, R.K.S., Kotasthane S.R., Khare M.N. and Gupta Om. 1985. Efficacy of seven methods in the detection of three *Fusarium* spp. associated with moong seeds. *Indian Journal of Mycology and Plant Pathology* 15 : 313-314.

7. Beniwal, S.P.S. and Chaubey S.N. 1979. Urdbean leaf crinkle virus : Effect on yield contributing factors, total yield and seed characters of urdbean. *Seed Research* 7 : 175-181.

8. Beniwal, S.P.S. and Chaubey S.N. 1984. Detection of urd bean leaf crinkle virus in urdbean seed. *Seed Research* 12 : 101-104.

9. Beniwal, S.P.S. and Chaubey S.N. 1984. Internal seed borne nature of urdbean leaf crinkle virus in urdbean seed. *Seed Research* 12 : 6-10.

10. Beniwal, S.P.S. Chaubey S.N. and Bharathan N. 1980. Presence of urdbean leaf crinkle virus in seeds of mung bean germplasm. *Indian Phytopathology* 33 : 360-361.

11. Beniwal, S.P.S. Chaubey S.N. and Matheswaran C. 1983. Some factors affecting transmission of urdbean leaf crinkle virus through seeds of urdbean (*Vigna mungo* (L). Hepper). *Seed Research* 11 : 95-99.

12. Bhargava, S.N. and Shukla D.N. 1979. Interrelationship studies on some pulses and oil crop seeds with their seed borne fungi. *Proceedings National Academy Sciences India* B, 49 : 81-84.

13. Bhikane, N.S. and Mukadam D.S. 1982. Impact of different conditions on the incidence of seed mycoflora of uridbean (*Phaseolus mungo* L.). *Indian Botanical Reporter* 1 : 55-57.

14. Bidari, V.B., Dayanand M. and Anahosur K.H. 1989. Field evaluation of seed treating fungicides against seed rot and poor stand in mungbean. *National Symposium New Frontiers of Pulses Research and Development, Kanpur*, 120 p.

15. Bilgrami, K.S., Prasad T., Jamaluddin and Roy A.K. 1976. Studies on the deterioration on some pulses by fungi. *Indian Phytopathology* 29 : 374-378.

16. Bilgrami, K.S., Sinha R.K. and Prasad T. 1978. Effect of the fungal flora on the seed content of mung. *Indian Phytopathology* 31 : 476-479.

17. Charya, M.A.S. and Reddy S.M. 1979. Seed mycoflora of mung (*Phaseolus aureus*) in Andhra Pradesh. *Maharashtra Vidnyan Mandir Patrika* 14(2) : 55-58.

18. Charya, M.A.S. and Reddy S.M. 1980. Effect of volatile and gaseous exudates of germinating seeds of mung (*Phaseolus aureus*) on spore germination of some seed borne fungi. *Biological Bulletin of India* 2(2) : 47-50.

19. Charya, M.A.S. and Reddy S.M. 1980. Production of lyases by *Phoma exigua* associated with seed rot of *Vigna radiata*. *Indian Journal of Mycology and Plant Pathology* 10: XXIII.

20. Charya, M.A.S. and Reddy S.M. 1980. Production of cell wall degrading enzymes by two seed borne fungi. *Current Science* 49 : 557-558.

21. Charya, M.A.S. and Reddy S.M. 1980. Ascorbic acid in germinating seeds of *Phaseolus aureus* Roxb. under pathogenesis. *Geobios* 7: 167-168.

22. Charya, M.A.S. and Reddy S.M. 1980. Effect of humidity on seed deterioration of mung (*Phaseolus aureus* Roxb.) by some seed borne fungi. *Maharashtra Vdnyan Mandir Patrika* 15(1) : 37-40.

23. Charya, M.A.S., Reddy S.M. and Reddy M.M. 1980. Evaluation of different volatile compounds against fungal decay of mung (*Vigna radiata*) seeds. *Maharashtra Vidnyan Mandir Patrika* 15(2) : 61-63.

24. Charya, M.A.S. and Reddy S.M. 1981. Deterioration of mung (*Vigna radiata* Wilczek) seeds due to certain seed borne fungi. *Indian Journal of Botony* 4: 80-82.

25. Charya, M.A.S., and Reddy S.M. 1981. Production of phosphatase and esterases in germinating mung seeds under pathogenesis. *Indian Phytopathology* 34 : 514-515.

26. Charya, M.A.S. and Reddy S.M. 1982. Toxic effect of *Fusarium oxysporum* on seed germination and growth of mung (*Vigna radiata*). *Indian Botanical Reporter* 1 :169-170.

27. Charya, M.A.S. and Reddy S.M. 1983. Effect of phenolic compounds and fungicides on cellulase (Cx) production by two seed borne fungi.*Indian Journal of Mycology and Plant Pathology* 13 : 337-339.

28. Charya, M.A.S., Reddy E.J.S. and Reddy S.M. 1983. Interaction of seed borne fungi and *Rhizobium* on nodulation and plant growth of mung *(Vigna radiata). Indian Journal of Mycology and Plant Pathology* 13 : 223-224.

29. Charya, M.A.S. and Reddy S.M. 1984. Inhibition of production of pectic enzymes of *Phoma exigua* seed rot fungus of *Vigna radiata. Indian Phytopathology* 37 : 537-539.

30. Charya, M.A.S. and Reddy S.M. 1984. Antifungal activity of some fungicides against four seed borne fungi. *Pesticides* 18(1) : 38-39.

31. Chaubey, Usha. 1978. Studies on seed borne fungi of mung with special reference to *Fusarium* species. M.Sc.Ag. Thesis J.N.K.V.V., Jabalpur, M.P. India, 96 pp.

32. Chindhalore, J.L. 1974. Studies on the role of seed borne pathogens of mung causing seed and seedling diseases and their control. M.Sc.Ag. Thesis, J.N.K.V.V., Jabalpur, M.P. India. 58 pp.

33. Chowdhary, A.K. and Nath P.S. 1983. A rapid method of inoculation of urdbean leaf crinkle virus (ULCV) using germinated seeds. *Indian Journal of Experimental Biology* 21: 158.

34. Gaur,R.B., Ahmed S.R. and Kataria P.K. 1984. Controlling *Xanthomonas phaseoli* (Smith) Dowson of mung seed through chemotherapy and heat therapy. *Hindustan Antibiotics Bulletin* 26 : 23-26.

35. Gupta, Pushpa Kumari and Gupta J.S.1981.Mycobial seed deterioration of *Phaseolus aureus* during storage. *International Symposium of Plant Pathology, IPS, New Delhi,* 217 p.

36. Gupta, Pushpa Kumari, Gupta J.S., Singh S. and Deo P.P. 1982. Studies on aflatoxin production by some new strains of *Aspergillus flavus* associated with stored seeds of mung *(Phaseolus aureus)* in Agra market. *Indian Journal of Mycology and Plant Pathology* 12 : 315.

37. Gupta, Pushpa Kumari and Gupta J.S. 1984. Storage deterioration of mungbean seed by fungi. *International Journal of Tropical Plant Diseases* 2 : 169-173.

38. Hussain, S.S. and Ahmad M.A. 1971. Studies on stored food grains fungi from pulses. *Pakistan Journal of Scientific and Industrial Research* 14 : 507-511.

39. ISTA. 1976. Proceedings of International Seed Testing Association. *Seed Science and Technology* 4 : 180 pp.

40. Jain, N.K. 1970. Studies on *Rhizoctonia bataticola* (Taub.)Butler causing diseases of urd *(Phaseolus mungo L.)*. M.Sc.Ag. Thesis, J.N.K.V.V., Jabalpur, M.P. India, 60 pp.

41. Jalali, B.L., Khirbat S.K., Sangwan M.S., Khurana A.L. and Dudeja S.S. 1981. Interaction of fungicidal treatment and Rhizobial inoculation on the growth and development of pulse crops. *Indian Phytopathology* 34 : 109.

42. Jamaluddin, Sinha R.K., Bilgrami K.S. and Prasad T. 1977. Changes in protein content of urid seeds (*Phaseolus mungo L.*) due to fungal flora. *Current Science* 46: 461.

43. Jharia, H.K. 1970. Studies on seed pathology of pulse crops important in Madhya Pradesh. M.Sc.Ag. Thesis, J.N.K.V.V., Jabalpur, M.P. India, 87 pp.

44. Jhooty, J.S. and Bains S.S. 1976. Modification of the toxicity of the benomyl by seed exudates of mung. *Indian Journal of Microbiology* 16: 73-75.

45. Kadian, O.P. 1981. Effect of some chemicals and heat on seed transmission of urdbean leaf crinkle virus. *Third International Symposium Plant Pathology, IPS, New Delhi,* 166-167 pp.

46. Kaiser, W.J. and Mossahebi G.H. 1974. Natural infection of mungbean by bean common mosaic virus. *Phytopathology* 64 : 1209-1214.

47. Kanta, S. 1982. Effect of culture filtrate of some fungi on seed germination and on seedlings of *Phaseolus mungo. Biological Bulletin of India* 4 : 78-84.

48. Kanta, S. and Vaidehi B.K. 1980. Effect of culture filtrate on the seed germination and on root and shoot elongation of *Phaseolus mungo. Indian Journal of Mycology and Plant Pathology* 10 : XXII.

49. Khare, M.N., 1974. Seed borne pathogens their detection and importance in relation to the need of quarantine regulations with special reference to legumes.*61st Indian Science Congress, Nagpur,* 2pp.

50. Khare, M.N. and Chaubey U. 1978. Significance, method of detection and control of five species of *Fusarium* associated with mung seeds. *National Academy of Sciences India,* 1p.

51. Khare, M.N., Mathur S.B. and Neergaard P. 1977. Seed borne fungi of pulses, their detection, location and importance. *18th International Seed Testing Association Congress, Medrid, Spain, Reprint No.80-S,* 3pp.

52. Kumar, C.S.K.V. 1981. *Cylindrocladium scoparium* Morg. a new record on blackgram seeds. *Current Science* 50 : 921-22.

53. Lassim, M.B.M., Chin, H.F. and Abdullah W.D. 1984. The effect of weathering on mungbean*(Vigna radiata (L.)*Wilczek) seed quality. *Pertanika* 7 : 77-81.

54. Montal, G.C., Nandi D. and Nandi B. 1985.Allyl-isothiocyanate as an effective post harvest preservative of seed. *Seed Science and Technology* 13 : 529-536.

55. Moore, E.S. 1931. Internal boll disease of cotton in South Africa. *South Africa Department of Science Bulletin* 94 :11-18.

56. Narayanaswamy, P. and Jaganathan T. 1975. Effect of black gram leaf crinkle virus infection on seed set and distribution of virus in seeds. *Madras Agricultural Journal* 62 : 151-154.

57. Nath, R., Mathur S.B. and Neergaard P. 1970. Seed borne fungi of mung.*(Phaseolus aureus* Roxb.) from India and their significance. *Proceedings International Seed Testing Association* 35 : 225-241.

58. Natrajan, T. and Bagyaraj D.J. 1981. Effect of fumigants on seed microflora of blackgram and field bean. *Indian Phytopathology* 34 : 512-514.

59. Natrajan, T. and Bagyaraj D.J. 1984. Fumigation effect on microflora and viability of blackgram and field bean seeds. *Pesticides* 18 (4) : 40-42.

60. Natrajan, T. and Bagyaraj D.J. 1985. Microflora associated with some pulse seeds. *Agricultural Digest* 5: 197-199.

61. Neergaard, P. 1977. "Seed Pathology", Vol. I, McMillon Press, London, 575 pp.

62. Paik, S.B. and Do. E.S. 1987. *Fusarium* sp. isolated from seed, root and cultivated soil of *Phaseolus* vi (ri) *dissimus* and their pathogenicity. *Korean Journal of Plant Pathology* 3: 8-12.

63. Phatak, H.C. 1974. Seed borne plant viruses identification and diagnosis and seed health testing. *Seed Science and Technology* 2 : 3-155.

64. Prasad, T., Bilgrami K.S., Roy A.K. and Jamaluddin. 1976. Deterioration of moong seeds by *Curvularia lunata. Seed Research* 4 : 191-193.

65. Providenti, R. 1986. Seed transmission of black eye cowpea mosaic virus in *Vigna mungo. Plant Disease* 70 : 981.

66. Punam, Kumari and Sinha K.K. 1986. Aflatoxin contamination in mung *(Vigna radiata). Indian Phytopathology* 39 : 146.

67. Punam,Kumari and Sinha K.K. 1987. Aflatoxin elaboration in mungbean seeds. *Indian Phytopathology* 40 : 201-203.

68. Ramachandran, P. and Chenulu V.V. 1985. Seed health testing for seed borne viruses-an urgent need. *Indian Journal of Plant Pathology* 3 : 33-37.

69. Rath, G.C. and Routray G.N. 1978. Internally seed transmitted diseases of mung and urid. *Science & Culture* 44 : 40-41.

70. Rashid, A.O.M.B., Barma A.C. and Shaikh M.A.Q. 1983. Seed borne fungi of mungbean and their pathogenicity. *Bangladesh Journal of Botany* 12 : 223-224.

71. Rayan, G.F. Greenblatt G. and Dolaimy K. 1961 Seedling albinism induced by an extract of *Alternaria tenuis. Science* 134 : 833-834.

72. Roy, A.N. and Sharma R.B., 1982. Effect of *Phaseolus mungo* L. seed coat leachate on fungal growth. *Phytopathologia Mediterranea* 21 : 31.

73. Saxena,R.M. and Gupta J.S. 1982. Effect of seed leachate on spore germintion of seed borne fungi of *Vigna radiata. Indian Phytopathology* 35 : 236-240.

74. Saxena, R.M. and Sinha S. 1977. Seed borne infection of *Vigna mungo* in Uttar Pradesh. *Indian Phytopathology* 30 : 582-583.

75. Saxena, R.M. and Sinha S. 1977. Seed borne infections on *Vigna radiata* (L.) Wilczek var. *radiata* in Uttar Pradesh-New records. *Science & Culture* 44 : 377-379.

76. Saxena, R.M. and Sinha S. 1979. Field mycoflora of *Vigna radiata* (L.) Wilczek var. *radiata* and *Vigna mungo* (L.) Hepper in relation to pre emergence and post emergence mortality. *Seed Research* 7 : 159-164.

77. Saxena, R.M. and Gupta J.S. 1979. Field fungi associated with seeds of *Vigna radiata* (L.) Wilczek var. *radiata* and *Vigna mungo* (L.) Hepper and their persistance during storage. *Proceedings of the Indian National Science Academy* B, 45 : 6 6-6 8.

78. Saxena, R.M. and Sinha S.1980. Storage mycoflora of *Vigna radiata* and *Vigna mungo* in relation to pre-emergence and post-emergence mortality. *indian Journal of Mycology and Plant Pathology* 10 : 120-121.

79. Sharda, R. and Shetty H.S. 1987. Location and transmission of *Macrophomina phaseolina* in blackgram seeds. *Indian Phytopathology* 40 : 194-196.

80. Sharma, H.C., Khare M.N., Chand J.N. and Kumar S.M. 1970. Efficacy of fungicides in the control of diseases of urid (*Phaseolus mungo L.). Recent Advances in Crop Production, U.P. Institute of Agricultural Sciences, Kanpur*, 271-274 pp.

81. Sharma, I. and Dubey G.S. 1981. Studies on seed borne nature of aphid transmitted urdbean leaf crinkle virus. *Third International Symposium on Plant Pathology, I.P.S., New Delhi*, 166 p.

82. Shukla, D.N. and Bhargava S.N. 1975. Some studies on *Fusarium oxysporum* Schl. ex Fries isolated from seeds of *Phaseolus mungo. Proceedings of the National Academy of Sciences India B*, 45 : 93-96.

83. Shukla, D.N. and Bhargava S.N. 1976. Fungi isolated from seeds of pulses. *Proceedings National Academy of Sciences India* B, 46 : 453-454.

84. Shukla, D.N. and Bhargava S.N. 1976. Some pathogenic fungi from pulses and oilseed crops. *Proceedings National Academy of Sciences, India B*, 46 : 531-532.

85. Singh, I. and Chohan J.S. 1973. Seed borne fungi of greengram (*Phaseolus aureus*) in the Punjab. *Indian Journal of Mycology and Plant Pathology* 3 : 199.

86. Singh, I. and chohan J.S. 1976. Seed borne fungi in blackgram (*Phaseolus mungo)* in the Punjab. *Indian Journal of mycology and Plant Pathology* 6 : 80.

87. Singh, V. 1977. Studies on *colletotrichum* Corda species associated with urd (*Phaseolus mungo*) seeds, M.Sc. Ag. Thesis, J.N.K.V.V., Jabalpur, M.P., India., 98 pp.

88. Sinha, R.K. 1979. Effect of seed borne fungi on protein contents of urad seeds (*Phaseolus mungo* L.) under different relative humidities. *Proceedings Third Botanical conference*, 18 pp.

89. Sinha, R.K. and Prasad T. 1981. Effect of fungal metabolites on seed germination, mycobial association and seedling growth of mung. *Indian Phytopathology* 34 : 515-517.

90. Sinha, R.K. and Roy A.K. 1981. Changes in seed contents of mung (*Vigna radiata*) due to *Aspergillus flavus* infestation at different relative humidities. *Indian Phytopathology* 34 : 119.

91. Soni, P.S. and Thind B.S. 1987. Detection of *Xanthomonas campestris* pv. *vignicola* (Burk.) Dye from cowpea seeds and *Xanthomonas campestris* pv. *vignaeradiatae* (Sabet *et al.*) Dye from mungbean seeds. *Phytopathologia Mediterranea* 26 : 1-6.

92. Sood, A.K., Thind B.S. and Singh N. 1976. Occurrence and control of bacterial leaf spot of mung (*Phaseolus aureus*) in the Punjab state. *Journal of Research Punjab Agricultural University* 13 : 182-185.

93. Suhag, L.S. 1975. Fungal flora of mung (*Phaseolus aureus*) seeds, pathology and control. *Indian Journal of Mycology and Plant Pathology* 5 : 165-168.

94. Suhag, L.S. and Suryanarayana D. 1976. Some aspects of seed health testing with respect to seed borne fungi of pulse crops grown in Haryana. *Indian Joural of Mycology and Plant Pathology*. 6 : 32-36.

95. Tandon, R.N. 1976. *Achaetomium*, a new record of seed borne pathogen of *Phaseolus aureus*. *Indian Phytopathology* 29 : 464-465.

96. Thind, B.S., Kumar Krishna and Soni P.S. 1984. Eradication of *Xanthomonas campestris* pv *vignaeradiatae* from mungbean seeds by physical and chemical treatments. *Indian Journal of Mycology and Plant Pathology* 14 : 8.

97. Varma, K., Sareen V.N.S. and Singh R. 1976. The effect of aflatoxin on allantoinase activity of germinating seeds. *Journal of Reasearch, Punjab Agricultural University* 13 : 99-104.

98. Vidhyasekaran, P. and Arjunan G. 1976. Fungicide treatment of blackgram seeds for the control of storage fungi. *Madras Agricultural Journal* 63 : 393-395.

99 Zainum, W. and Hasbullah M. 1982. Storage of mungbean seed (*Vigna radiata* (L.) Wilczek) inoculated with four species of *Aspergilli*. *Pertanika* 5 : 212-218.

100. Zote, K.K. and Mayee C.D. 1982. Influence of fungicidal seed treatment during storage on seed borne fungi of mungbean. *Pesticides* 16 (4) : 10-12.

CHAPTER 7
NON PARASITIC DISORDERS

With the introduction of new agricultural technology,use of agrochemicals has increased. It has resulted in several new problems in crop production in the form of phytotoxicity or the defficiency symptoms on crop plants. Advancement in the industries has created the pollution problems. Blackgram and greengram are subjected to these problems resulting in reduced yield.

7.1. Injuries due to agrochemicals :

7.1.1. Heavy metals : Heavy metals are ingredients of some of the common fungicides like organomercurial compounds. It has been reported that mung bean seedlings when treated with heavy metals like mercury and lead, the δ-aminolevulinic acid hydrolase activity is decreased ultimately reducing total chlorophyll contents. The enzyme activity is localized exclusively in chloroplasts of the plant cells (10).

7.1.2. Oxycarboxin (Plantvax) : It is a systemic fungicide. When used as seed treatment, it significantly reduces the reducing, non reducing, total water soluble sugars and the true protein contents of blackgram seed. It is reported to delay the seed germination also to a considerable extent (3). However, nitrogen content of leaves of plants raised from plantvax treated seed is not affected (2). The phosphorus contents of such plants are also reduced but sulphur and magnesium contents are increased considerably. The nitrogen content of leaves is decreased if they are sprayed with the fungicide.

7.1.3. Benomyl (Benlate) : It is a systemic fungicide and is translocated in mungbean and urdbean plants from roots to foliage and the movement is faster in mung bean as compared to urd bean (5). Treatment of greengram and blackgram seeds with this fungicide inhibits the germination, radicle growth and the allantoinase activity in seed (8).

7.1.4. Carboxin (Vitavax) : It is also reported to inhibit the germination, radicle growth and allantoinase activity in the blackgram and greengram seed (8).

7.1.5. Actidione : As reported in case of benomyl and vitavax, actidione when used as seed treatment of greengram and blackgram also inhibits the germination, growth of the radicle and the activity of allantoinase in the seed (8). It is more effective than benomyl and vitavax.

7.1.6. Bavistin (carbendazim) : Like cytokinins, the bavistin is believed to change the cell permeability and thus affect the enzyme activity. The leaves of blackgram incubated in the solution of bavistin showed a dose dependent fall in the nitrate reductase. This effect was observed in case of detached nodules also. The enzyme activity increased up to 1 % concentration of bavistin but decreased at the higher concentrations (1). In case of RU 2 urdbean variety, author has observed phytotoxic effect of 0.1 % concentration of bavistin, derosol and Topsin M (all carbendazim) when used as spray (unpublished data). The leaves became yellow resulting in early and forced maturity and reduced yield and plant growth. This observation is based on one season data only and needs further confirmation.

7.1.7. 2, 4-D injury : Accidental spray of the herbicide 2, 4-D is reported to cause wide spread injuries on pulse crops including urd bean chickpea and lentil. It caused varying symptoms of deshaping of leaves to the appearance of warty blisters on shoot and roots (6)

7.2. Injuries due to pollution :

7.2.1. Petrocoke dust : Petrocoke dust is the particullate pollutant of the petrolium refineries. When mungbean plants were treated with petrocoke at 2 g/m^2 day for 40 consecutive days between 26 and 65 days of plant age, the plants exhibitted reduction in root and shoot length and in numbers of leaves, nodules, flowers and pods of the plant (9).

7.2.2. Sulphur dioxide and NO_2 : Sulphur dioxide is the important polluting agent in the smoke of industrial centres and is highly injurious to crop plants (4). It is a colourless gas with a characteristic suffocating odor and has bleaching effect on organic colouring matters like chlorophyll and pigments of flowers. Both are always present in the air in small amount but if their quantity increases, they adversely affect the crop growth and yield.

The SO_2 is reported to reduce the yield and shoot root weight of mungbean. Its effect is correlated with its dose. A higher cumulative dose appreciably reduces the growth and productivity of mungbean plants (11). It is more injurious when present alone while in combination with NO_2 it is less injurious to mungbean plants (7). The injurious effect of NO_2 alone is insignificant.

REFERENCES

1. Chandra, G. and Mathur S.N. 1986. Responses of *Vigna mungo* (L.) Hepper to treatment of bavistin : Effect of nitrogen utilizing enzymes. *Pesticides* 20 (4) : 24-26.

2. Chopra, S.L. and Pruthi S.P. 1971. Effect of plantvax on the chemical composition of black gram (*Phaseolus mungo* Roxb.). *Indian Joural of Agricultural Sciences* 41 : 345-349.

3. Chopra, S.L. and Pruthi S.P. 1973. Effect of plantvax (2, 3-dihydro-5-carboxinilido-6-mehtyl 1,4-oxathiin 4, 4-dioxide) on the germination of black gram (*Phaseolus mungo*). *Journal of Researach Punjab Agriculture University* 10 : 86-90.

4. Heald, F.D. 1933. "Manual of Plant Diseases". *McGraw Hill Book Company Inc. New York,* 953 pp.

5. Lucose, C. and Singh R.D. 1983. Absorption and translocation of benomyl in different parts of some plants. *Indian Journal of Mycology and Plant Pathology* 13 : 326-328.

6. Nene, Y.L. 1969. Recent occurrence of some non infectious disease problems in the Nainital Tarai. *Indian Phytopathology* 22 : 164.

7. Pande, P.C. 1985. Effects of growth and photosynthesis of *Vigna radiata* cv. Pusa Baisakhi in environment containing SO_2 and/or NO_2. *Current Science* 54 : 393-394.

8. Paul, B., Thapar V.K. and Singh R. 1974. Effect of systemic fungitoxicants on metabolism in germinating seeds. *Indian Journal of Agricultural Sciences* 43 : 662-666.

9. Prasad, B.J. and Rao D.N. 1981. Growth responses of *Phaseolus aureus* plants to petrocoke pollution. *Journal of Experimental Botany* 32 : 1343-1350.

10. Prasad, D.D.K. and Prasad A.R.K. 1987. Effect of lead and mercury on chlorophyll synthesis in mung bean seedlings *Phytochemistry* 26 : 881-883.

11. Singh, N. and Rao D.N. 1986. Influence of sulphur dioxide on the growth and productivity of *Phaseolus aureus* plants. *Acta Botanica Indica* 14 : 230-235.

SUBJECT INDEX

E